Springer Theses

Recognizing Outstanding Ph.D. Research

Aims and Scope

The series "Springer Theses" brings together a selection of the very best Ph.D. theses from around the world and across the physical sciences. Nominated and endorsed by two recognized specialists, each published volume has been selected for its scientific excellence and the high impact of its contents for the pertinent field of research. For greater accessibility to non-specialists, the published versions include an extended introduction, as well as a foreword by the student's supervisor explaining the special relevance of the work for the field. As a whole, the series will provide a valuable resource both for newcomers to the research fields described, and for other scientists seeking detailed background information on special questions. Finally, it provides an accredited documentation of the valuable contributions made by today's younger generation of scientists.

Theses are accepted into the series by invited nomination only and must fulfill all of the following criteria

- They must be written in good English.
- The topic should fall within the confines of Chemistry, Physics, Earth Sciences, Engineering and related interdisciplinary fields such as Materials, Nanoscience, Chemical Engineering, Complex Systems and Biophysics.
- The work reported in the thesis must represent a significant scientific advance.
- If the thesis includes previously published material, permission to reproduce this must be gained from the respective copyright holder.
- They must have been examined and passed during the 12 months prior to nomination.
- Each thesis should include a foreword by the supervisor outlining the significance of its content.
- The theses should have a clearly defined structure including an introduction accessible to scientists not expert in that particular field.

More information about this series at http://www.springer.com/series/8790

Assim Boukhayma

Ultra Low Noise CMOS Image Sensors

Doctoral Thesis accepted by
École Polytechnique Fédérale de Lausanne, Switzerland

Springer

Author
Dr. Assim Boukhayma
STI IMT ICLAB
École Polytechnique Fédérale de Lausanne
Neuchâtel 2
Switzerland

Supervisor
Prof. Christian Enz
STI IMT ICLAB
École Polytechnique Fédérale de Lausanne
Neuchâtel 2
Switzerland

ISSN 2190-5053 ISSN 2190-5061 (electronic)
Springer Theses
ISBN 978-3-319-68773-5 ISBN 978-3-319-68774-2 (eBook)
https://doi.org/10.1007/978-3-319-68774-2

Library of Congress Control Number: 2017956744

Printed on acid-free paper

This Springer imprint is published by Springer Nature
The registered company is Springer International Publishing AG
The registered company address is: Gewerbestrasse 11, 6330 Cham, Switzerland

When a man dies, his acts come to an end, but three, ongoing charity, or knowledge by which people benefit, or a righteous son.

Muhammad ibn Abdullah (570–632 CE)

Supervisor's Foreword

I have known Assim Boukhayma since October 2012 when CEA-LETI asked me whether I would agree to be the advisor of his Ph.D. work focused on low-noise CMOS imagers for low-light vision. He was finishing his masters thesis at CEA-LETI during which he worked on the design of CMOS Terahertz image sensors. At this time, I thought that it would be more appropriate to follow up with a Ph.D. based on the results he obtained during his master in the field of Terahertz imaging. Assim finally convinced me that there was actually an increased interest for low-light vision offering new opportunities to develop extremely low-noise CMOS imagers. He actually was right.

Indeed, we very quickly found out that despite the many industrial groups working on the topic, none had really looked in details on the overall noise of CMOS imagers, from the simple four-transistor pixel to the complete read-out chain. After performing a detailed noise analysis of the complete read-out chain, we realized that there were still some degrees of freedom that could be used to minimize the noise despite the very constrained design. Indeed, the pixel circuit only includes four transistors. Assim designed a new pixel array in a standard CMOS image sensor (CIS) process that demonstrated an input-referred noise less than 0.5 electrons rms. Following this breakthrough, he then designed a full VGA CMOS imager that proved the concept to be working in a real array. These outstanding results were actually achieved only by circuit optimization using a standard pinned photodiode in the pixel. We then looked at how low-noise CMOS imagers could take advantage of technology scaling and showed that the gate current and its inherent shot noise started to limit the noise reduction. We showed that the optimum technology would be around 90 nm or 65 nm to bring the noise below 0.3 electrons rms.

Assim did his Ph.D. in a truly international context and had at the same time an experience in academia, industry, and government institutions learning about research methodology, technology transfer, and the strategic aspect of research. During his Ph.D., Assim worked on the design of Ultra Low Noise CMOS Image Sensors under my guidance and with the support of CEA-Leti. He proved his autonomy and showed a good working methodology from the analytical problem formalization to the proposition of conclusive solutions. Through the implementation of his solution, he experienced and developed skills in the design, layout, packaging, PCB design prototyping, and characterization of image sensors. His efforts led to three patents and several publications. All the chips he designed were tested successfully and achieved world record low noise levels. Besides from the academic sphere, his work also led to a successful technology transfer to industry and great interest among the community of image sensors professionals and enthusiasts.

Neuchâtel 2, Switzerland
Mai 2017

Christian Enz

Parts of this thesis have been published in the following articles:

- A. Boukhayma, A. Peizerat and C. Enz. A Sub-0.5 Electron Read Noise VGA Image Sensor in a Standard CMOS Process, in IEEE Journal of Solid-State Circuits (JSSC), vol. 51, pp. 2180–2191, 2016.
- A. Boukhayma, A. Peizerat and C. Enz. Noise Reduction Techniques and Scaling Effects towards Photon Counting CMOS Image Sensors, in Sensors, 16(4), 514, 2016.
- A. Boukhayma, A. Dupret, J.P. Rostaing and Christian Enz. A Low-Noise CMOS THz Imager Based on Source Modulation and an In-Pixel High-Q Passive Switched-Capacitor N-Path Filter, in Sensors, 16(3), 325, 2016.
- A. Boukhayma, A. Peizerat and C. Enz. Temporal Readout Noise Analysis and Reduction Techniques for Low-Light CMOS Image Sensors, in IEEE Transaction on Electron Devices (TED), vol. 63, pp. 72–78, 2016.
- A. Boukhayma, A. Dupret and C. Enz. A noise reduction circuit technique for CMOS terahertz imaging, in the International Conference on Noise and Fluctuations (ICNF), Xian, 2015.
- C. Enz and A. Boukhayma. Recent trends in low-frequency noise reduction techniques for integrated circuits, in the International Conference on Noise and Fluctuations (ICNF), 2015 International Conference on, Xian, China, 2015.
- C. Enz, F. Krummenacher and A. Boukhayma. Simple thermal noise estimation of OTA-based switched-capacitor filters, in the International Conference on Noise and Fluctuations (ICNF), Xian, China, 2015.
- A. Boukhayma, A. Peizerat and C. Enz. A correlated multiple sampling passive switched capacitor circuit for low light CMOS image sensors, in the International Conference on Noise and Fluctuations (ICNF), Xian, China, 2015.
- A. Boukhayma and C. Enz. A new method for kTC noise analysis in periodic passive switched-capacitor networks, in the IEEE New Circuits and Systems Conference (NEWCAS), Grenoble, France, 2015.
- A. Boukhayma, A. Peizerat and C. Enz. A 0.4 e-rms Temporal Readout Noise 7.5 μm Pitch and a 66% Fill Factor Pixel for Low Light CMOS Image Sensors, in the International Image Sensors Workshop (IISW), Vaals, Nederlands, 2015.
- A. Boukhayma and C. Enz. Design optimization for low light CMOS image sensors readout chain, in the IEEE New Circuits and Systems Conference (NEWCAS), Trois-Rivieres, QC, Canada, 2014.
- A. Boukhayma et al. A 533pW NEP 31$\times$31 pixel THz image sensor based on in-pixel demodulation, in the European Solid State Circuits Conference (ESSCIRC), 2014, Venice Lido, Italy, 2014.
- A. Boukhayma A. Peizerat, A. Dupret and C. Enz. Comparison of two optimized readout chains for low light CIS, in SPIE Proceedings, vol. 9022, Image Sensors and Imaging Systems, 2014.

Contents

1 Introduction 1
1.1 A Short Historical Review 1
1.2 Modern Solid State Imaging 3
1.3 Motivation 5
1.4 Main Contributions 7
1.5 Thesis Outline 9
References 10

2 Low-Noise CMOS Image Sensors 13
2.1 Pinned Photodiodes 13
2.1.1 Brief Historical Review 13
2.1.2 Device Physics 15
2.1.3 Device Operation 20
2.2 CIS Global Architecture 22
2.3 Pixel Architectures 23
2.3.1 Source Follower Based Pixel 24
2.3.2 Common Source Based Pixel 25
2.4 Column-Level Amplification 26
2.5 Correlated Sampling and Analog-to-Digital Conversion 28
2.5.1 Correlated Double Sampling 28
2.5.2 Correlated Multiple Sampling 29
2.6 Summary 32
References 33

3 Noise Sources and Mechanisms in CIS 35
3.1 Photon Shot Noise 35
3.2 Dark Current 36
3.2.1 Depleted Area Generation 38
3.2.2 Field-Free Area Generation 39
3.2.3 Dark Current Shot Noise 40

3.3 Transfer Noise 40
3.3.1 Charge Transfer Non-idealities 40
3.4 Electronic Noise 42
3.4.1 Thermal Noise 44
3.4.2 1/f and RTS Noise 47
3.4.3 Leakage Current Shot Noise 51
3.5 Fixed Pattern Noise 54
3.5.1 Spatial Gain Variation 55
3.5.2 Offset Variations 56
3.6 Discussion 57
References 59

4 Detailed Noise Analysis in Low-Noise CMOS Image Sensors 61
4.1 Preamble 61
4.2 Noise Calculation Method 63
4.3 Impact of CMS on Noise 65
4.4 In-Pixel Source Follower Based Readout Chain 67
4.4.1 The Conversion Gain 67
4.4.2 Thermal Noise 69
4.4.3 $1/f$ Noise 73
4.4.4 Leakage Current Shot Noise 74
4.5 In-Pixel Common Source Based Readout Chain 76
4.5.1 Conversion Gain 76
4.5.2 Thermal Noise 78
4.5.3 $1/f$ Noise 79
4.5.4 Leakage Current Shot Noise 80
4.6 Discussion 80
References 82

5 Noise Reduction in CIS Readout Chains 85
5.1 Thermal Noise Reduction 85
5.1.1 Column-Level Circuits Optimization 86
5.1.2 Pixel-Level Optimization 88
5.2 Flicker Noise Reduction 88
5.2.1 Correlated Sampling 89
5.2.2 Pixel-Level Optimization 90
5.3 Thin Oxide Source Follower: A Good Match 95
5.4 Summary 98
References 98

6 Design of a Sub-electron Readout Noise Pixel in a Standard CIS Process 101
6.1 The Optimized Pixel Design 101
6.2 Test Chip Architecture 102
6.3 Test Description and Measurement Results 104

6.3.1 Test Setup 104
6.3.2 Conversion Gain Measurement 106
6.3.3 Input-Referred Read Noise 107
6.3.4 Pixel Voltage Swing 108
6.4 Conclusion 110
Reference 110

7 Characterization of a Sub-electron Readout Noise VGA Imager in a Standard CIS Process 111
7.1 Chip Architecture and Circuit Design 111
7.1.1 Imager Architecture 111
7.1.2 Pixel Design and Layout 112
7.1.3 Column-Level Amplifier 113
7.1.4 Column-Level SSADC 115
7.1.5 Physical Implementation 117
7.2 Test and Characterization 118
7.2.1 Experimental Setup 118
7.2.2 Conversion Gain Measurement 119
7.2.3 Temporal Read Noise 120
7.2.4 Photo-Response Non-uniformity 121
7.2.5 Dark Current 121
7.2.6 Imager Lag 122
7.2.7 Quantum Efficiency 123
7.2.8 Imaging Demonstration 123
7.3 Discussion and Comparison to State-of-the-Art 125
7.4 Conclusion 127
References 127

8 A Passive Switched-Capacitor Circuit for Correlated Multiple Sampling 129
8.1 Introduction 129
8.2 Motivation 129
8.3 Design Principle 130
8.4 Implementation 131
8.5 Transient Noise Simulations 133
8.6 Conclusion 136
References 136

9 Downscaling Effects Towards Photon Counting Capability in CIS 137
9.1 Photon and Electron Count Conditions 138
9.2 Impact of Technology Downscaling on the Read Noise 139
9.2.1 $1/f$ and Thermal Noise 139
9.2.2 Leakage Current Shot Noise 141
9.2.3 Random Telegraph Signal 142

9.3 Summary . . . 142
References . . . 143

10 An Ultra Low Noise CMOS THz Imager . . . 145
10.1 Introduction . . . 145
10.2 Operation of CMOS THz Imagers . . . 146
10.3 Noise Reduction Mechanism . . . 147
10.4 Architecture of the Low Noise CMOS THz Imager . . . 148
10.4.1 Overall Architecture . . . 148
10.4.2 In-Pixel Antenna and Rectifier . . . 148
10.4.3 In-Pixel Amplification . . . 150
10.4.4 In-Pixel Filtering . . . 151
10.5 Analysis of the High-Q Filter . . . 152
10.5.1 Passive SC N-Path Filter . . . 152
10.5.2 Optimization with a G_m-C Filter . . . 155
10.6 Circuit Implementation . . . 157
10.7 Test and Characterization . . . 158
10.7.1 Baseband Characterization . . . 158
10.7.2 THz Characterization . . . 160
10.8 Summary . . . 164
References . . . 165

11 Conclusion . . . 167
11.1 Read Noise Reduction in CMOS Image Sensors . . . 167
11.1.1 General Methodology for Read Noise Reduction in CIS . . . 167
11.1.2 The Transistor Oxide Thickness: A Degree of Freedom for Noise Optimization . . . 169
11.1.3 Thin Oxide Source Follower Implementation for Sub-0.5-electrons Noise in a Standard CIS Process . . . 169
11.1.4 A New CMS Circuit Design for Further Noise Reduction . . . 170
11.1.5 Technology Downscale: Lower 1/*f* and Thermal Noise but Higher Leakage Shot Noise . . . 171
11.2 Read Noise Reduction in Terahertz CMOS Image Sensors . . . 171
11.3 Future Work . . . 172
11.3.1 CMOS Image Sensors . . . 172
11.3.2 Terahertz CMOS Image Sensors . . . 172
References . . . 173

Appendix A: Signal Variance Conservation with Sampling . . . 175

Author Biography . . . 179

Chapter 1
Introduction

This Chapter begins by a short historical overview paying tribute to some ground breaking works that shaped the science of optics and photography. A brief concentrated state of the art recalling the main technological device-level trends in modern image sensors and sensitive solid state visible photons detectors are summarized. The motivations behind this work are pointed out, the main contributions are summarized and the organisation of this manuscript is presented.

1.1 A Short Historical Review

In the year 1021, Alhazen Ibn-Alhaytham publishes *Kitab al-manazir*: the book of optics. This early scientific work brought ground breaking knowledge to the field of optics. Alhazen was the first to present the images perceived by the human eye as the result of the light reflection by an infinite number of points from the scene objects. He gave the name of *primary* light to the source of light and *secondary* light to the one reflected by the non-self-luminous bodies. He claimed that the light originating from the *primary* sources propagates in straight lines. Based on his theoretical and experimental findings, he established the reflection and refraction laws and derived the magnification formulas of lenses. Alhazen was also the first to analyse the anatomy of the human eye based on geometrical optics. His works inspired him to invent the *camera obscura* which is the ancestor of today's cameras [1, 2].

It was later, in the XIXth century that paper and plates covered with a thin layer of photosensitive material like silver chloride were used in order to capture the image created in the *camera obscura*. This process was further developed to give birth to photographic films. By the end of the XIXth century, Kodak, by George Eastman, was the first commercially successful consumer film camera. Film technology created and supplied the market of photography for decades until the late nineties of the XXth century [3].

A. Boukhayma, *Ultra Low Noise CMOS Image Sensors*,
Springer Theses, https://doi.org/10.1007/978-3-319-68774-2_1

Fig. 1.1 Summarized timeline of the history of photography

Modern physics changed our perception of light and matter at the microscopic scale. The pioneering works of Albert Einstein and Max Plank revealed the quantized nature of light. Albert Einstein discovered a fundamental aspect of light which is the photo electric effect. This principle was the first step and still the background of electronic imaging.

Modern physics also led to the development of semiconductors which led to the invention of the charge coupled device (CCD) in 1969 by Willard Boyle and George E. Smith at Bell labs that was initially meant to be used as an electronic shift register [4]. Silicon was the substrate of choice for semiconductor technology and CCD in particular. By chance, the band-gap energy of silicon is exactly what it is supposed to be in order to absorb a visible light photon and generate an electron within the silicon. It did not take much time to Michael F. Tompsett before he sees the potential of the CCD device for imaging applications [5]. This was the beginning of a new era of electronic imaging which is not only a new era of photography but the beginning of the age where it became possible for light information to be captured, stored, and analyzed electronically (Fig. 1.1).

1.2 Modern Solid State Imaging

In parallel to the CCD development, complementary metal oxide semiconductor (CMOS) technology was constantly improving in terms of performance and miniaturization. By the early nineties, it was possible to include several transistors in one pixel while still achieving acceptable pitch. The first electronic pixels including the sense node and electronic amplification appeared in the nineties. The CMOS image sensors (CIS) started by occupying the market of low cost and low performance image sensors and took advantage of the mobile electronic devices (smartphones, tablets, digital cameras...) exploding markets. Quickly, CMOS image sensors became the technology of choice with respect to speed, resolution, power consumption and on-chip integration.

For several years, signal to noise ratio (SNR) used to be the performance wall preventing CIS from invading the small niche markets left to CCDs. The continuous improvements of CIS, during the first decade of this century, led to the introduction and maturation of pinned photodiodes (PPDs) in CIS. The barrier of one photoelectron noise performance have been crossed using process modification [7] and even in standard CIS technology [8, 9]. Process level optimization of CIS can today lead to a dynamic range over a hundred dB [10]. Back side illumination together with microlens and color filter layers became standards. Figure 1.2 shows three cross sections of CIS chips from different manufacturers achieved by chipworks reprinted and adapted from [6]. This figure shows the fruit of several years of development driven by the huge handset and smartphone market demand. Three important technological advances are featured by Fig. 1.2. Figure 1.2a shows the 2011 generation CIS chips from Samsung where one can distinguish the microlens layer on top of the color filters and metal layers which distinguish a front-side illuminated CIS.

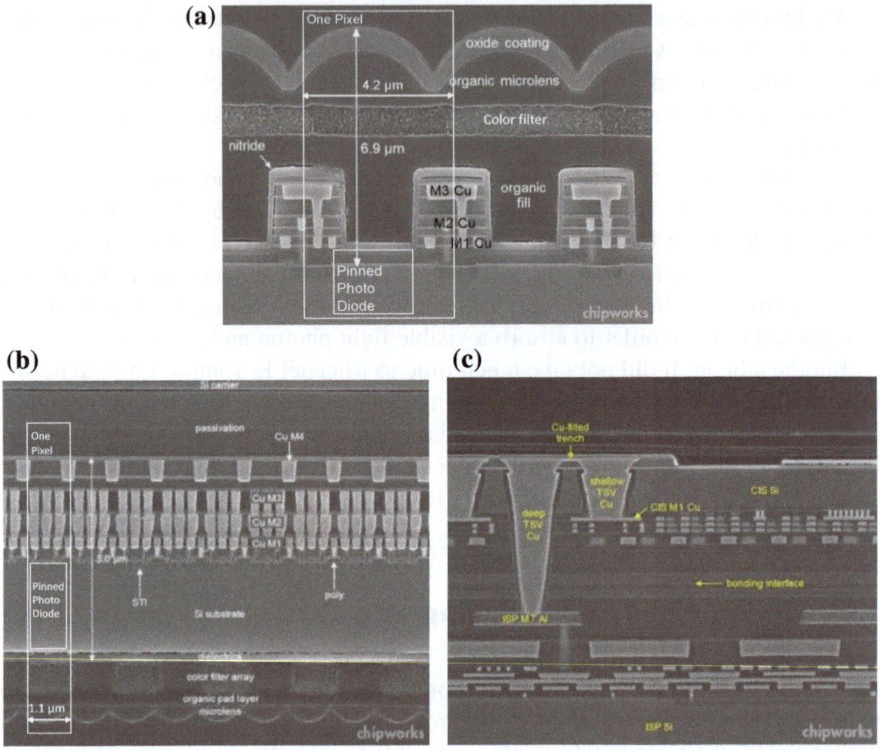

Fig. 1.2 Three CIS chips cross sections: **a** from the Samsung NX200 showing an example of front-side illuminated CIS **b** from OmniVision OmniBSI-2 showing an example of back-side illuminated CIS and **c** Sony showing an example of a stacked CIS. The figures are reprinted and adapted from [6]

Figure 1.2b shows a more recent generation chip from Omnivision illustrating the implementation of back-side illumination in CIS. In this example the transistors with metal layers and microlenses with color filters are in two opposite sides of the chip. In this way, the quantum efficiency is significantly improved since metal layers and in-pixel transistors do not interact with the incident light but rather reflect back part of the light not absorbed by silicon to the PPD. The third example in Fig. 1.2c shows the Sony most recent stacked technology. In this case, a back side illuminated CIS chip is stacked with another chip dedicated for the digital processing. The two chips metal layers are connected with deep Through-silicon vias (TSVs). In this way, the pixel analog circuitry and logic circuitry can be separated, not only in two chips, but also in two different technology nodes.

1.3 Motivation

As mentioned above, electronic imaging applications go beyond photography. Low light vision, scientific, space and medical imaging as well as photon counting for scientific instrumentation and industrial applications emerged with electronic imaging. All these niche applications require an extremely low noise performance from the devices targeting these markets. Figure 1.3 shows different visible light sensor technologies compared in terms of noise performance. The CIS based on pixels with 3 transistors (3T) and a photodiode without a transfer gate are the devices featuring the highest read noise. The read noise in 3T CIS is dominated by the reset noise sampled at the sense node after the reset resulting in a noise floor standard deviation generally of about a few tens of electrons (e^{-}_{rms}). The CCDs offer a lower noise performance of a few e^{-}_{rms} thanks to the double sampling readout scheme that cancels the reset noise mentioned above and reduces the electronic low frequency noise. Pixels based on PPDs are referred to as 4T pixels for having the additional transfer gate separating the sense node from the photodiode with respect to 3T pixels. The 4T CIS reproduce the same readout process used at the output of the CCD chips but at the pixel level. Commercial 4T CIS feature a noise level comparable to CCDs [11], but the versions optimized for low noise can reach the one e^{-}_{rms} limit. With a read noise below 0.3 e^{-}_{rms}, it becomes possible to count the photoelectrons with an accuracy of 90%. The only commercial solid state semiconductor devices reaching this limit are electron multiplication CCDs (EMCCDs) [12], avalanche photodiodes (APDs) and single photon avalanche photo diodes (SPADs) [13]. All these devices introduce amplification at the early stage of the photoelectron generation by applying an electric field high enough to accelerate the photoelectron in order to multiply by

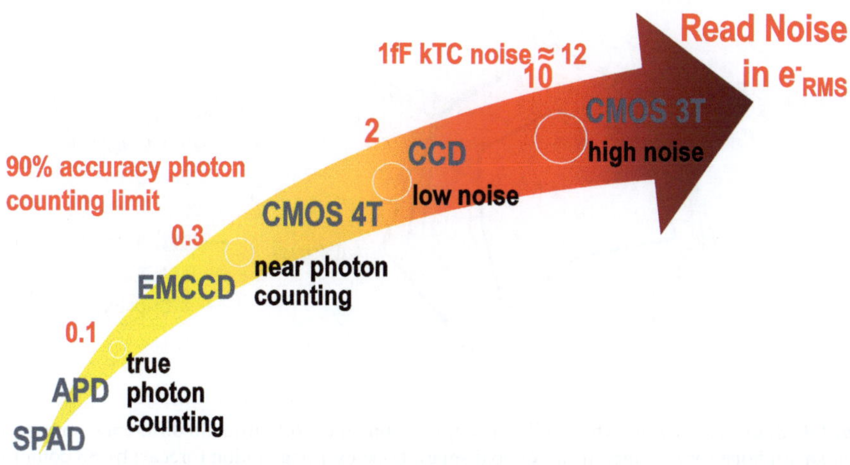

Fig. 1.3 Visible light solid state sensor technologies classified in terms of the read noise

impact ionisation or trigger an avalanche effect. This high electric field is also one of the drawbacks of such devices precluding their large scale integration.

Engineering is all about continuous improvement towards ultimate performance. Further improving the noise performance of CIS can lead to deep subelectron noise performance which will allow photoelectron counting. The first logical implication is that the same low-cost, low-power, on-chip uncooled CIS imager that one could have in his smartphone would be capable of the same performance than SPADs and EMCCDs while delivering high quality images. This means, in the long term, the emergence of a new range of applications besides high quality imaging or even new imaging paradigms like quanta image sensors [14]. The implication in the short term is that CIS mobile cameras will be capable of delivering shot noise limited images. Theoretically, smartphones would allow visibility where the naked eye can't see.

In fact, the first motivation for extremely sensitive CIS is directly related to the market demand and particularly smartphone consumer expectations. Figure 1.4a shows statistics reporting how smartphones owners use their devices based on a population

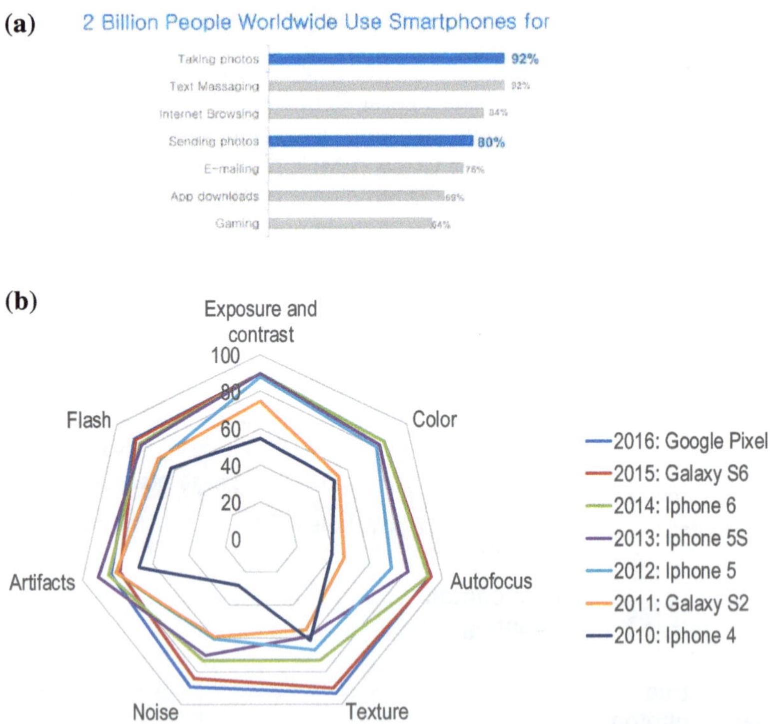

Fig. 1.4 Statistics showing what 2 billion people around the world from 88 different countries use their smartphone for reprinted from: Global smartphone user penetration forecast by 88 countries, 2014, Strategy Analytics **a**. Smartphone camera scores from [15] showing the detailed scores of the top smartphone of the year between 2010 and 2016, pointing out the efforts made in the noise performance enhancement

of 2 billion people in more than 88 countries around the world. Taking photos and sharing them figure on the top of the users smartphone utility. This fact makes the image quality at the center of the market battlefield between smartphone manufacturers as well as image sensors design houses. Figure 1.4 shows how low noise is becoming one of the most important criteria of smartphone benchmarks. It shows image sensors performance scores achieved by smartphones that recorded the highest global scores between 2010 and 2016 based on [15]. Figure 1.4 illustrates the amount of effort put in each point improving the image quality in smartphones and it appears clearly that the low-light performance and noise in mobile CIS is becoming one of the major concerns.

The motivation of this work is driven by all these trends and market opportunities. But on the top of all of that, manipulating a few photon or handling a single electron is, from a scientific perspective, a very exciting experience. An opportunity to go into the deep meanings and implications of the microscopic physical models of semiconductors as well as the quantized nature of light and charge.

1.4 Main Contributions

The main contributions of this thesis are summarized in the following items.

- A review of CMOS image sensors based on PPDs is presented in Chap. 2. Starting with a historical background, a summary of the PDDs physics is reviewed. The in-pixel readout circuit topologies exploiting the PPD are discussed fundamentally and the overall architecture of classical low noise CISs is presented.
- The physical mechanisms behind the random fluctuations affecting the signal at different levels of conventional CIS readout chains are reviewed and clarified in Chap. 3. These include the photon shot noise, the dark current shot noise, the charge transfer nonidentities and the electronic circuits noise. Practical examples from measurements illustrating the different noise sources are presented.
- This thesis dedicates a particular focus to the readout circuit $1/f$ and thermal noise given that these noises preclude the ultimate limit of photoelectron counting in conventional CIS. In Chap. 4, a detailed analytical calculation of $1/f$ and thermal noise in readout chains based on the two possible in-pixel configurations is presented, namely the in-pixel source follower and common source topologies. The detailed analysis reveals process and design level noise reduction techniques. These are studied analytically and based on simulation results in Chap. 5.
- Among the noise reduction techniques suggested by the analytical noise calculation, the increase of the oxide capacitance by using a thin oxide in-pixel amplifying transistor, for low $1/f$ noise, is revealed for the first time. A readout chain design based on a thin oxide PMOS source follower is presented in Chap. 6. The in-pixel source follower is optimally sized thanks to the analytical expression of the input referred $1/f$ noise mentioned above. In order to validate the low noise performance of the newly proposed design. A test chip has been designed and fabricated in a

180 nm CIS process. It embeds small arrays of the proposed new pixels together with state-of-the-art 4T pixels based on buried channel source followers optimized at process level for low $1/f$ noise. The new pixels feature a pitch of 7.5 µm and a fill factor of 66%. A mean input-referred noise of 0.4 e^-_{rms} is reported, for the first time, with a pixel designed in a standard CIS process. Compared with the state-of-the-art pixels, also present onto the test chip, the mean rms noise is divided by more than 2.

- Based on these encouraging result, a full VGA (640 H × 480 V) imager has been integrated in a standard 180 nm 4PM CIS process. The presented imager relies on a 4T pixel of 6.5 µm pitch with a properly sized and biased thin oxide PMOS source follower. A full characterization of the proposed image sensor, at room temperature, is presented in Chap. 7. With a pixel bias of 1.5 µA the sensor chip features an input-referred noise histogram from 0.25 e^-_{rms} to a few e^-_{rms} peaking at 0.48 e^-_{rms}. This sub-0.5 electron noise performance is obtained with a full well capacity of 6400 e^- and a frame rate that can go up to 80 fps. The VGA imager also features a fixed pattern noise as low as 0.77%, a lag of 0.1% and a dark current of 5.6 e^-/s. It is also shown that the implementation of the in-pixel n-well does not impact the quantum efficiency of the pinned photo-diode.
- Correlated multiple sampling (CMS) is another noise reduction technique commonly used in low noise CIS to reduce thermal noise by averaging and slightly further reduce $1/f$ noise compared to a simple correlated double sampling (CDS). A passive switched-capacitor network, with a minimum number of capacitors, performing fast CMS is introduced in Chap. 8. The proposed circuit requires no additional active circuitry, has no impact on the output dynamic range and does not need multiple analog-to-digital conversions. The new CMS circuit is verified with transient noise simulations and shows a noise reduction in perfect agreement with the ideal CMS.
- This paper also discusses the impact of technology scaling on the CIS read noise. It shows how designers can take advantage of scaling and how the MOS transistor gate leakage tunneling current appears as a challenging limitation. For this purpose, both simulation results of the gate leakage current and $1/f$ noise data reported from different foundries and technology nodes are used.
- Enhancing the sensitivity by minimizing the electronic readout noise through circuit techniques can also benefit to CMOS image sensors capturing electromagnetic radiations outside the visible range like Terahertz (THz) imagers. THz imaging is an emerging technology with various possible applications in security and non-invasive sensing. Micro-bolometers have been so far the technology of choice for building focal plane THz arrays. But given the high cost of this hybrid technology, THz imaging, in a CMOS process appears as a promising alternative. CMOS THz detectors use metal antennas coupled to MOS transistors used as rectifiers. These detectors suffer from a lower sensitivity compared to micro-bolometers. One way to increase the sensitivity is by reducing the readout chain noise. Unlike visible light imaging, THz imaging is active. This implies that the THz source can be controlled and synchronized with the sensor chip. This additional degree of freedom can be exploited for noise reduction. In this context, the first integration of

a 1 kpixel CMOS THz imager with in-pixel high-selective filtering synchronized with the THz source modulation is presented. This noise reduction technique acts as chopper-stabilized amplifiers. The integration, in each pixel, of a metal antenna with a MOS transistor rectifier, low noise amplification and highly selective filtering, based on a switch-capacitor N-path filter combined with a broad band Gm-C filter, has been successfully tested. The measured Q factor is 100 for a modulation frequency of a few hundreds of kHz. The measured input-referred noise of the readout chain is as low as 0.2 μV RMS corresponding to a total noise equivalent THz power of 0.6 nW at 270 GHz and 0.8 nW at 600 GHz.

1.5 Thesis Outline

The next chapters of this thesis are organized as follows.

- Chapter 2 presents a review of CMOS image sensors based on PPDs. Starting with a historical background, a summary of the PDDs physics is reviewed. The in-pixel readout circuit topologies exploiting the PPD are discussed and the overall architecture of classical low noise CISs is presented.
- Chapter 3 is dedicated to the physical mechanisms behind the random fluctuations affecting the signal at different levels of conventional CIS readout chains. These include the photon shot noises, the dark current shot noise, the charge transfer non-idealities and the electronic circuits noise. Practical examples from measurements illustrating the different noise sources are presented.
- Chapter 4 puts the focus on the readout circuit $1/f$, thermal and shot noise in conventional CIS readout chains. A detailed analytical noise calculation for the two possible in-pixel configurations is presented, namely the in-pixel source follower and common source topologies.
- Chapter 5 shows how the latter detailed analysis reveals process and design level noise reduction techniques. These are studied analytically and based on simulation results. Among the noise reduction techniques suggested by the analytical noise calculation, the increase of the oxide capacitance by using a thin oxide in-pixel amplifying transistor, for low $1/f$ noise, is revealed for the first time.
- Chapter 6 presents a readout chain design based on a thin oxide PMOS source follower and implementing the noise reduction techniques presented in the previous chapters. It also presents the test chip designed in a 180 nm CIS process and embedding small arrays of the proposed new pixels together with state-of-the-art 4T pixels based on buried channel source followers optimized at process level for low $1/f$ noise. This Chapter presents experimental results verifying the theoretical expectations presented above.
- Based on these primary encouraging result, Chap. 7 presents a full VGA (640 H × 480 V) imager integration in a standard 180 nm 4PM CIS process and demonstrates how a careful design based the presented noise calculation has led to an input-referred noise histogram from 0.25 e^-_{rms} to a few e^-_{rms} and peaking at 0.48 e^-_{rms}.

A full characterization of the VGA imager is also presented including photon transfer curves, noise, pixel-to-pixel non-uniformity, lag, dark current and quantum efficiency.

- Chapter 8 presents another contribution for low noise CIS readout chain design. A passive switched-capacitor network, with a minimum number of capacitors, performing fast correlated multiple sampling (CMS) is introduced. The proposed circuit requires no additional active circuitry, has no impact on the output dynamic range and does not need multiple analog-to-digital conversions. The new CMS circuit is verified with transient noise simulations and shows a noise reduction in perfect agreement with ideal CMS.
- Chapter 9 discusses the impact of the technology downscale on CIS sensitivity from an electronic read noise perspective. It shows promising opportunities for $1/f$ and thermal noise but points out threads regarding random telegraph signal (RTS) noise and leakage current shot noise.
- Chapter 10 presents another noise reduction technique in the emerging field of terahertz imaging. In this example, the additional degree of freedom consisting in the active terahertz source control is exploited. This chapter presents the first integration of a 1 kpixel CMOS THz imager with in-pixel high-selective filtering synchronized with source modulation. The pixel design is based on a passive switched capacitor N-path filter combined with a Gm-C filter.

References

1. A.M. Smith, Getting the big picture in perspectivist optics. Isis **72**(4), 568–589 (1981)
2. D. Lindberg, *Theories of Vision from Al-kindi to Kepler*. Chicago History of Science and Medicine (University of Chicago Press, Chicago, 1981), pp. 58–86
3. History of kodak, http://www.kodak.com/US/en/corp/kodakHistory/. Accessed 29 Apr 2016
4. W. Boyle, G. Smith, Buried channel charge coupled devices, Patent US 3,792,322 (1974), https://www.google.com/patents/US3792322
5. M. Tompsett, Charge transfer imaging devices, Patent US 4,085,456 (1978), https://www.google.com/patents/US4085456
6. R. Fontaine, Innovative technology elements for large and small pixel cis devices, in *International Image Sensors Workshop* (2013)
7. Y. Chen, Y. Xu, Y. Chae, A. Mierop, X. Wang, A. Theuwissen, A 0.7e-rms temporal-readout-noise cmos image sensor for low-light-level imaging, in *2012 IEEE International Solid-State Circuits Conference Digest of Technical Papers (ISSCC)* (2012), pp. 384 –386
8. C. Lotto, P. Seitz, T. Baechler, A sub-electron readout noise cmos image sensor with pixel-level open-loop voltage amplification, in *2011 IEEE International Solid-State Circuits Conference* (2011), pp. 402–404
9. A. Boukhayma, A. Peizerat, C. Enz, Temporal readout noise analysis and reduction techniques for low-light cmos image sensors. IEEE Trans. Electron Devices **63**(1), 72–78 (2016)
10. S. Wakashima, F. Kusuhara, R. Kuroda, S. Sugawa, A linear response single exposure cmos image sensor with 0.5e- readout noise and 76ke- full well capacity, in *2015 Symposium on VLSI Circuits (VLSI Circuits)* (2015), pp. C88–C89
11. M.H. White, D.R. Lampe, F.C. Blaha, I.A. Mack, Characterization of surface channel ccd image arrays at low light levels. IEEE J. Solid-State Circuits **9**(1), 1–12 (1974)

12. M.S. Robbins, B.J. Hadwen, The noise performance of electron multiplying charge-coupled devices. IEEE Trans. Electron Devices **50**(5), 1227–1232 (2003)
13. E. Charbon, Single-photon imaging in complementary metal oxide semiconductor processes. Philos. Trans. R. Soc. Lond. A Math. Phys. Eng. Sci. **372**, 2014 (2012)
14. E. Fossum, The quanta image sensor (qis): concepts and challenges, in *Imaging and Applied Optics* (Optical Society of America, 2011), p. JTuE1, http://www.osapublishing.org/abstract.cfm?URI=COSI-2011-JTuE1
15. Dxomark, https://www.dxomark.com/. Accessed 29 Apr 2016

Chapter 2
Low-Noise CMOS Image Sensors

CMOS image sensors (CIS) as known today are the fruit of several decades of research and development culmination starting from the discovery of the photodetecting effect of pn junctions and the implementation of charge coupled devices (CCDs) as image sensors to the development of pinned photo-diodes (PPDs) in CMOS technology. Today, CIS is a mature technology, PPDs became standard devices and the optimal readout circuit schemes for each application are becoming well known. This Chapter recalls the historical background of CIS, presents a simplified physical model of the PPD and reviews the readout chain architectures implemented in the low-noise context.

2.1 Pinned Photodiodes

The pinned photo diode (PPD) is a photosensitive structure at the heart of almost all CMOS image sensors and even CCDs. PPDs became the first choice device in solid state image sensors due to their CMOS compatibility, high quantum efficiency and low dark current in addition to a readout process based on correlated double sampling which is prone to a low noise readout by canceling the reset noise and mitigating low frequency noise.

2.1.1 Brief Historical Review

The idea of using silicon devices as photodetectors goes back to 1965 when the photosensing effect of pn junctions has been revealed for the first time by G.P. Weckler [1, 2]. These early works showed that a reverse biased pn junction behaves as a capacitor charged by a photocurrent proportional to the incident light intensity.

A. Boukhayma, *Ultra Low Noise CMOS Image Sensors*,
Springer Theses, https://doi.org/10.1007/978-3-319-68774-2_2

This device became the key element in the first MOS passive pixel sensors (PPS) designed by P.J.W. Noble [3] in 1968. These pn junctions were also introduced in the CCDs in a device called Inter Line Transfer (ILT) [4] in order to avoid using the CCD structure for both the light integration and the charge transfer. A transfer gate was introduced, in each pixel, between the pn junction and the CCD structure transferring the integrated photoelectrons through the whole column to the sense node located at the output of the chip for the readout. But the pn junction photodiodes performance was mainly limited by the sampled noise after reset, commonly referred to as the kTC noise and the incomplete charge transfer during the readout [5]. To address these issues, N. Teranishi invented the buried pn junction that was first presented in 1982 [6]. This buried structure consisted in adding a heavily doped p+ thin layer on the top of the n layer of the pn junction making it a vertical p+np structure. This device had also another virtue consisting in an extremely lower dark current compared to pn junctions [6]. The name of PPD has first been given to this structure in a 1984 publication [7] presenting an enhancement of the buried photo diode quantum efficiency through the reduction of the p+ layer thickness.

Thanks to the miniaturization of CMOS devices, the integration of in-pixel amplification became possible. E.R. Fossum was the first to take advantage of this aspect. He proposed the in-pixel integration of charge transfer and source follower buffering in 1993 [8, 9] and gave it the name of active pixel sensor (APS). This CMOS structure used a photogate as a detection device. The implementation of PPDs in CMOS pixels required further technological improvements to allow the charge transfer from the PPD to the sense node with low voltages compatible with CMOS processes. Such improvements emerged in the late 90 [10, 11] and early 2000 [12, 13]. The name of 4 T pixels is commonly given to the pixel structure combining the APS topology with a PPD phototdetector. In other words, an APS where the photogate is replaced by a PPD. The name of 4 T pixels was given by contrast to the 3 T pixel structure designating a pixel with a photodiode directly connected to a source follower with a reset and selection switches. The 4 T pixel has the additional transfer gate separating the PPD from the sense node. The fabrication of APS with PPDs in CMOS processes was a major turning point in the image sensor industry. From this point, the CMOS image sensors started challenging and advancing CCDs even at the sensitivity level. The sensitivity of CIS based on PPD was further optimized in terms of quantum efficiency by introducing back-side-illumination (BSI). BSI was first introduced for CCDs to enhance their quantum efficiency by exposing the back side of the chip to the incident light instead of the front side which is partially covered by the gates and metal wires [14, 15]. BSI was adapted to the CIS industry and became a standard in the late 2000 included even in the mass produced consumer applications [5]. The color filter and micro lenses layers are stacked in the back side of the chip and the front side metal layers act as light reflector further increasing the quantum efficiency.

2.1.2 Device Physics

As mentioned in the historical review, the PPDs have been first developed in CCD technology for their low spill back and lag, low dark current and good quantum efficiency. These devices have been implemented after in CIS. The PPD structure consists in a np junction buried under a shallow highly doped p+ thin layer. A few physical models of the PPD have been proposed in the literature [16–18]. In this section, we give a basic physical model of the PPD leading to a simple derivation of the potential shape and pin voltage value and position. The PPD is nothing else than two pn junctions sharing the same n doped area as depicted in Fig. 2.1. The junctions give rise to depletions in both sides of each junction. When the concentration of free carriers is mainly given by the doping concentration, the full depletion approximation can be used. The latter assumes that the depletion length in each side of the junctions have a clear edge and the transition between the depleted (charged) and non-depleted (neutral) regions is abrupt [19]. These assumptions are justified by the exponential dependence of the carrier concentration on the gap between the Fermi level and band edges.

Consider the case when the depletion regions of the p+n and np junctions merge in the n layer. The n layer of the PPD becomes fully depleted as depicted in Fig. 2.1. x_n marks the edge between the p+ and n layers while x_p marks the border between the latter and the epitaxial p bulk. x_{p_1} is the depletion edge abscissa of the p+n junction and x_{p_2} corresponds the depletion edge for the p junction. The charge distribution, under the full-depletion approximation, corresponding to the structure described in Fig. 2.1 is shown in Fig. 2.2a. For the sake of simplification, we do not consider the case of linear doping concentration in the n layer.

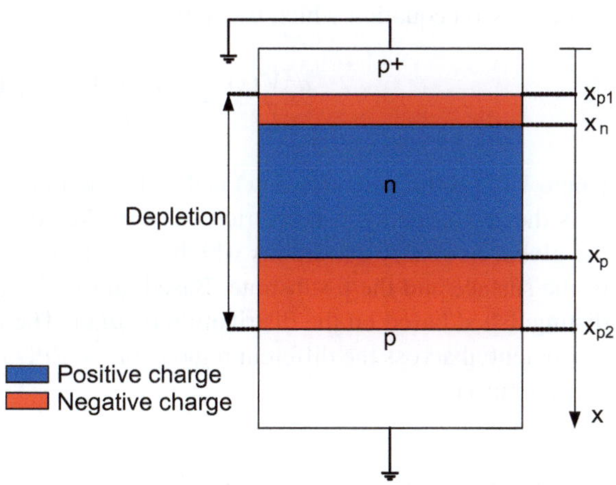

Fig. 2.1 Structure of the stacked p+np layers in the PPD

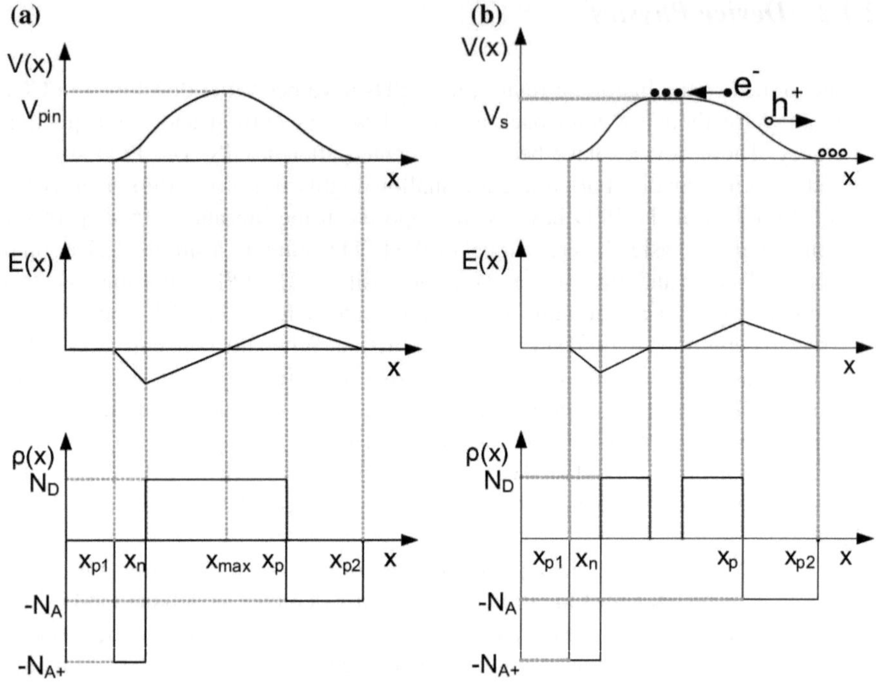

Fig. 2.2 Charge density, electric field and potential under the full depletion approximation for a fully depleted PPD (**a**) and after integration (**b**)

The key equation relating the electric field and potential to the charge distribution is the Poisson equation which is expressed as:

$$\frac{d^2V(x)}{dx^2} = -\frac{dE(x)}{dx} = -\frac{\rho(x)}{\varepsilon_s}, \tag{2.1}$$

where $V(x)$ is the potential, $E(x)$ is the electric field, $\rho(x)$ is the charge density and ε_s is the semiconductor dielectric constant. We add to these differential equations the global boundary conditions which are field free and zero potential at the level of the surface and the p substrate. Based on Fig. 2.2a, five different regions can be distinguished based on the distribution of $\rho(x)$. The continuity of the electric field and potential across the different regions of the PPD is also a constraint to take into consideration.

2.1.2.1 Neutral p+ and p Regions

The neutral p+ region is delimited by $0 \le x < x_{p_1}$. For the p epitaxial substrate it corresponds to $x_{p2} \le x$. These areas of the semiconductor are not depleted, hence the charge density is nul ($\rho(x) = 0$). Consequently, these two areas are field free with a zero potential.

2.1.2.2 Depleted p+ Region

The depleted region of the p+ layer is located within $x_{p_1} \leq x < x_n$. The charge density in this area is determined by the hole impurities density as

$$\rho(x) = -qN_{A^+}. \tag{2.2}$$

N_{A^+} is of the order of 10^{18} to $10^{19}\,\mathrm{cm}^{-3}$ which is one to two orders of magnitude higher than the concentration of donors in the n layer and a few orders of magnitude higher that the p substrate concentration ranging between 10^{15} and 10^{16}.

The electric field is then derived by solving (2.1) for the boundary condition of $E(x_{p_1}) = 0$ verifying the continuity of the electric field at x_{p_1}. $E(x)$ simplifies to

$$E(x) = -\frac{qN_{A^+}}{\varepsilon_s}(x - x_{p_1})\,\mathrm{for}\, x_{p_1} \leq x < x_n. \tag{2.3}$$

Consequently, the potential over x in this region can be expressed, in the way that verifies its continuity at x_{p_1}, as

$$V(x) = \frac{qN_{A^+}}{2\varepsilon_s}(x - x_{p_1})^2\,\mathrm{for}\, x_{p_1} \leq x < x_n. \tag{2.4}$$

2.1.2.3 Depleted n Region

This region corresponds to $x_n \leq x < x_p$. The depleted n region is in a certain way the result of the merging between the depletions of the p+n and np junctions. The charge density under full depletion is given by the density of donor impurities N_D. The density of donors in this area is considered to decrease linearly with x. Here we suppose N_D constant for simplification since this linear dependence over x does not have a fundamental impact on the potential shape. The charge density is hence expressed as

$$\rho(x) = qN_D. \tag{2.5}$$

The electric field is then expressed using the Poisson equation (2.1) and the continuity condition at $x = x_n$ using (2.3). This yields to

$$E(x) = \frac{qN_D}{\varepsilon_s}(x - x_n) - \frac{qN_{A^+}}{\varepsilon_s}(x_n - x_{p_1})\,\mathrm{for}\, x_n \leq x < x_p. \tag{2.6}$$

Consequently, the potential verifying the Poisson equation and the continuity at x_n can be expressed as

$$V(x) = -\frac{qN_D}{2\varepsilon_s}(x - x_n)^2 + \frac{qN_{A^+}}{\varepsilon_s}(x_n - x_{p_1})x - q\frac{qN_{A^+}}{\varepsilon_s}(x_n - x_{p_1})(x_n + x_{p_1}) \text{ for } x_n \leq x < x_p. \tag{2.7}$$

2.1.2.4 Depleted p Region

The depleted region of p substrate corresponds to $x_p \leq x < x_{p_2}$. The charge density in this area is determined by the hole impurities density in the epitaxial substrate as

$$\rho(x) = -qN_A. \tag{2.8}$$

The electric field is then derived by solving (2.1) for the boundary condition of $E(x_{p_2}) = 0$ verifying the continuity of the electric field at x_{p_2}. $E(x)$ simplifies to

$$E(x) = -\frac{qN_A}{\varepsilon_s}(x - x_{p_2}) \text{ for } x_p \leq x < x_{p_2}. \tag{2.9}$$

Consequently, the potential over x in this region can be expressed, in the way that verifies is continuity at x_{p_2}, as

$$V(x) = \frac{qN_{A^+}}{2\varepsilon_s}(x - x_{p_2})^2 \text{ for } x_p \leq x < x_{p_2}. \tag{2.10}$$

Note that the continuity must also be verified between the depleted p and n regions. This condition leads to the following equation

$$N_A x_{p_2} - N_{A^+} x_{p_1} = (N_A + N_D)x_p - (N_{A^+} + N_D)x_n. \tag{2.11}$$

This equation sets the relationship between the doping concentrations and the depletion width of the PPD.

2.1.2.5 Pin Voltage

Based on the Eqs. (2.4), (2.7) and (2.10), the potential takes its maximum in the depleted n region. The position x_{max} corresponding to this maximum corresponds to the abscissa where the electric field is nul. Hence x_{max} is obtained by solving the equation $E(x) = 0$ in the depleted n region as

$$x_{max} = x_n + \frac{N_{A^+}}{N_D}(x_n - x_{p_1}). \tag{2.12}$$

Let us first verify that x_{max} is comprised between x_n and x_p. It is clear from (2.13) that x_{max} is higher than x_n. The continuity condition (2.11) can be used to express x_{max} as a function of x_p. Based on this equation x_{max} can also be expressed as

$$x_{max} = x_p - \frac{N_A}{N_D}(x_{p_2} - x_p). \tag{2.13}$$

Hence x_{max} is comprised between x_n and x_p.

The pin voltage can be expressed by substituting x_{max} in (2.7) as

$$V_{pin} = \frac{qN_{A^+}}{2\varepsilon_s} \cdot \frac{N_{A^+} + N_D}{N_D}(x_n - x_{p_1})^2. \tag{2.14}$$

The pin voltage expression obtained in (2.14) is exactly the built-in voltage of the p+n junction. Indeed this result shows that the pin voltage of the p+np structure of the PPD can be determined by analyzing the Fermi level gaps between the layers. The Fermi level gap between the p+ and n layers is given by

$$V_{bi1} = \frac{kT}{q} ln\left(\frac{N_{A^+}N_D}{n_i^2}\right), \tag{2.15}$$

where V_{bi1} is the built-in potential of the p+n junction. The Fermi level gap between the n layer and the epitaxial p is given by

$$qV_{bi2} = \frac{kT}{q} ln\left(\frac{N_A N_D}{n_i^2}\right), \tag{2.16}$$

where V_{bi2} is the built-in potential of the pn junction. As mentioned previously, the doping concentration in the p+ layer is a few orders of magnitude higher than the one in the epitaxial p layer. Hence

$$V_{bi1} > V_{bi2} \tag{2.17}$$

Consequently, the built-in potential in the p+np structure when the Fermi levels of the different layers become equal is given by the highest energy gap which corresponds to the p+n junction. Finally the pin voltage can be expressed as

$$V_{pin} = \frac{kT}{q} ln\left(\frac{N_{A^+}N_D}{n_i^2}\right). \tag{2.18}$$

For instance, for a doping concentration in the p+ layer N_{A^+} of $10^{18}\,cm^{-3}$ and a donors concentration in the n layer of $10^{16}\,cm^{-3}$, the pin voltage at 300 K would be 0.8 V.

Note that in the simplified analysis presented above, the linear doping concentration over the PPD n well [5] is not considered and replaced by a flat concentration over the n well. This analysis does not include the horizontal border effects on the PPD and the impact of the PPD width and length on the pin voltage [20].

2.1.2.6 Charge Photo-Generation and Integration

The band gap energy of the silicon is about 1.1 eV. A visible photon has an energy ranging between 3.5 and 1.5 eV. For front side illumination, the thin p+ layer on the top of the PPD is transparent and the photon gets absorbed by silicon atoms located in the depleted region. Since the photon energy is higher than the silicon band gap, an electron-hole pair is generated. Due to the electric field in the depleted region (Fig. 2.2a), the electron is attracted by the position corresponding to the maximum potential and the hole sinks in the ground through the substrate. Each photoelectron located at the depleted n well compensates the positive charge of a donor hole. Hence, with cumulated photoelectrons, a neutral region starts to grow in both sides of the maximum voltage position (x_{max}) as depicted by Fig. 2.2b. The PPD is saturated when the depleted region in the n well disappears due to the cumulated electrons. Consider a PPD section of area A. The total positive charge in the PPD volume corresponding to the section A in the n well between x and x_{max} is given by $q \cdot N_D \cdot A \cdot (x_{max} - x_n)$. Hence, a number of $N_D \cdot A \cdot (x_{max} - x_n)$ photoelectrons is enough to saturate this section of a PPD, since there will be no maximum voltage, which gives an estimation of the PPD full well capacity FWC. For a doping concentration of $10^{16}\,\text{cm}^{-3}$ and an n layer depth of 0.25 μm the rough estimation of the maximum storage charge is about $2.5\,\text{ke}^-/\mu\text{m}^2$.

2.1.3 *Device Operation*

Figure 2.3 shows the conventional schematic of a pixel based on a PPD. The cross section of a PPD allows to see stacked n and p+ layers as well as the shallow trench insulator (STI) used to isolate the PPD. A heavily doped p+ area separates the STI oxide from the PPD n well and p epitaxial layer in order to reduce the impact of interface imperfections as will be shown in the next Chapter. The transfer gate (TG) is used to control the potential barrier at the edge of the PPD. When the TG voltage is low enough, the potential under the transfer gate is lower enough than the pin voltage of the PPD in order to keep the integrated charge in the PPD n well. Typical values of the TG low voltage range between 0 V and slightly negative values for dark current issues as will be discussed in the next Chapter. The sense node is simply an n+p junction capacitance. The voltage at the surface of the n + layer is set to a high voltage in the 2.5–3.3 V range creating a depletion at the n+p interface. The sense node voltage is read by an adequate electronic circuit that will be detailed in the next Section. After sensing the reset level, the TG voltage is increased creating a depletion

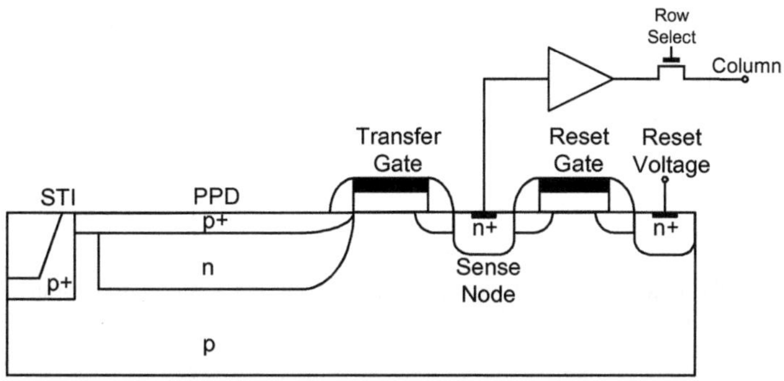

Fig. 2.3 Global architecture of a pixel based on a PPD

under the transfer gate where the potential is higher that the PPD pin voltage and in the mean time lower than the sense node maximum voltage. During this time, the depletion regions under the sense node, the transfer gate and in the PPD n well merge and the electrons cumulated in the PPD during the integration phase diffuse towards the higher potential area under the transfer gate and then to the sense node where the potential is even higher. The charge diffusion to the sense node causes its voltage level to drop from the reset level with a step proportional to the transferred charge. This voltage is sensed and the reset level voltage is then subtracted to obtain the signal level only assuming the reset level has remaind constant. This operation is called correlated double sampling (CDS). It not only subtracts the constant reset level but also reduces the noise that is correlated between the reset time and the transfer time such as the flicker noise [21].

The charge storage and transfer discussed above can be described by the hydraulic model of Fig. 2.5 that shows the different readout steps according to the operation timing diagram of Fig. 2.4 showing also the reset and transfer samples of the CDS.

Fig. 2.4 Readout timing diagram of a conventual pixel based on a PPD

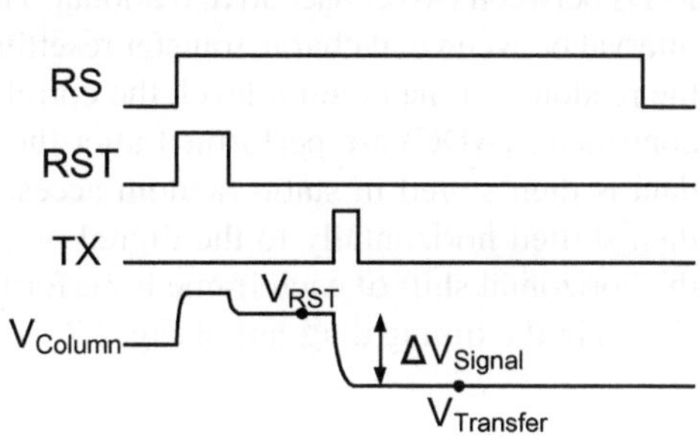

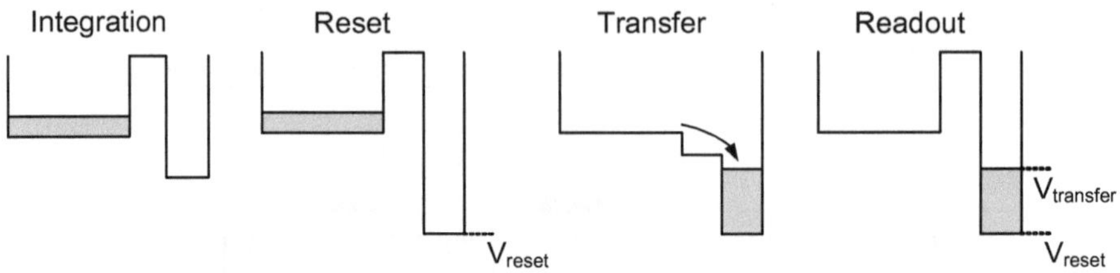

Fig. 2.5 Hydraulic model of a PPD with the transfer gate and sense node regions showing the different readout steps

2.2 CIS Global Architecture

Figure 2.6 shows the overall block diagram of a conventional low noise CIS. The pixels array is at the center of the imager. It occupies most of the chip silicon area. Each pixel comprises a pinned photodiode with at least one amplifying transistor and three MOS switches for reset, transfer and row selection. In order to achieve high frame rates, a column parallel readout scheme is generally performed. The pixels array is read line by line and all the pixels of the same line are read in parallel. Hence, at the top of the lines, a mixed signal control block made of shift registers and level shifters is generally implemented in order to drive the pixel lines by generating the row selection, reset and transfer commands. At the bottom of the columns, analog amplification is generally implemented before the correlated sampling and analog-to-digital conversion. Figure 2.7 shows the timing diagram of a conventional CIS readout chain. It shows the main line control signals as well as the timing of the column level signal processing. Each pixel line is generally addressed with the same frequency as the frame rate. Hence the PPDs remain exposed to the light between two consecutive readouts. In order to control the exposure time of the pixels, an intermediate reset and transfer operation can be performed in order to empty the PPDs between two consecutive readouts. The integration time is then set by the time interval between that charge transfer resetting the PPD and the one performed during the readout. At the column level, the correlated sampling and the analog-to-digital-conversion (ADC) are performed after the column-level amplification. The digital data is then stored in static random access memories SRAMs. These SRAMs are then shifted horizontally to the digital output of the chip. For a higher frame rate, the horizontal shift of each frame is performed at the beginning of the next frame as shown in the timing diagram of Fig. 2.7.

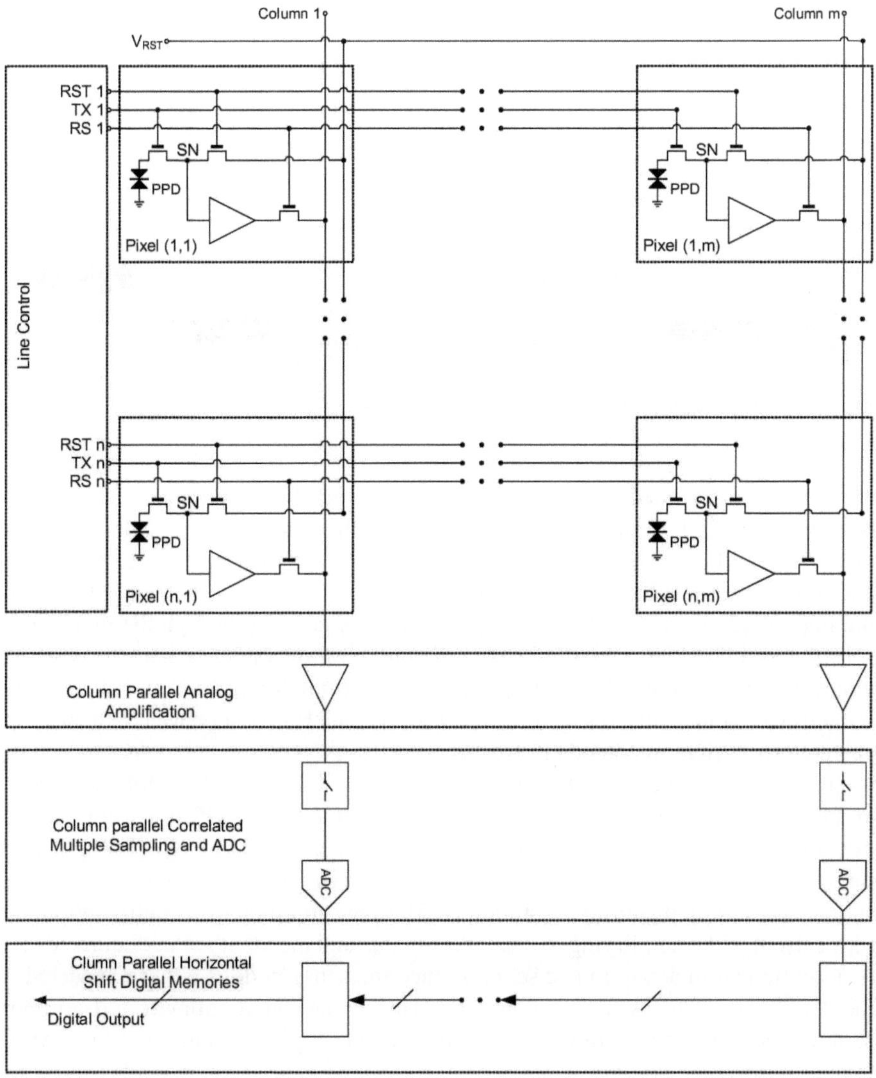

Fig. 2.6 Block diagram of a low noise CIS

2.3 Pixel Architectures

In any sensor readout chain, the low noise amplification must be applied at the earliest stage of the readout chain. In the case of CCDs, the amplification was applied at the output level of the chip since the CCDs do not offer the possibility of integrating any electronics at the pixel level [22]. Indeed the charge was transferred from pixel to pixel vertically then horizontally until the sense node located at the chip output. The

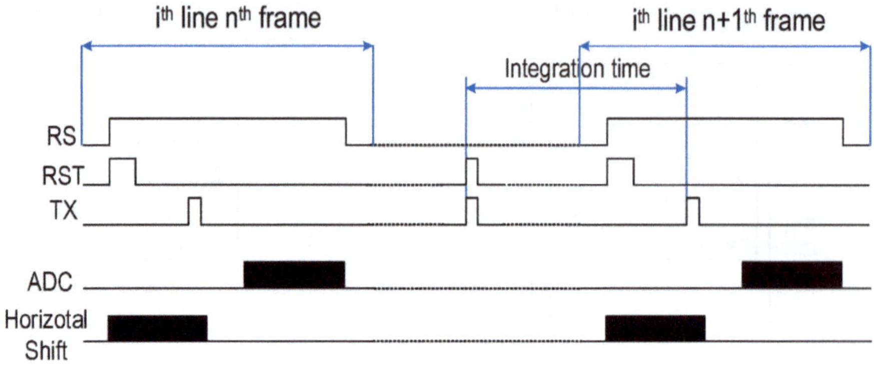

Fig. 2.7 Timing diagram of a conventional low noise CIS

idea of integrating electronic amplification at the pixel level goes back to the early beginnings of CMOS image sensors [23, 24] with 3 T pixels. The main advantage of this in-pixel amplification is the separation between the sense node of the pixel and the column level parasitic capacitance. Indeed, without in-pixel amplification, the photodiode would be directly connected to the column through the pixel (row) selection switch MOS transistor [2]. In this configuration, the sense node capacitance would be of the order of several pF. Hence, each photoelectron integrated in the photodiode would only result in a few 160 nV which is a very weak signal to be detected by CMOS uncooled electronics.

In order to optimize the quantum efficiency and fill factor of the pixel, the in-pixel amplification is introduced with the minimum number of MOS transistors. When operated in saturation, a single MOS transistor can operate as a voltage amplifier in three possible topologies, namely, the source follower (common drain), the common source and common gate configurations [25]. In the common gate configuration, the bias current of the amplifying transistor flows through the input voltage source which is not practical in active image sensors since the sense node is either connected to the photodiode or to a sense node junction. Hence, the source follower and common source topologies are the most popular in-pixel amplifying schemes used in CMOS image sensors.

2.3.1 Source Follower Based Pixel

The in-pixel source follower (SF) topology is the most common configuration in CIS. Figure 2.8a shows the schematic of the source follower stage. The sense node is connected to the gate of the amplifying transistor and the output voltage is produced at its source. In this configuration, the transistor is used as a voltage buffer. The output voltage as a function of the sense node voltage is depicted in Fig. 2.8b. This DC characteristic shows two important advantages of the source follower scheme.

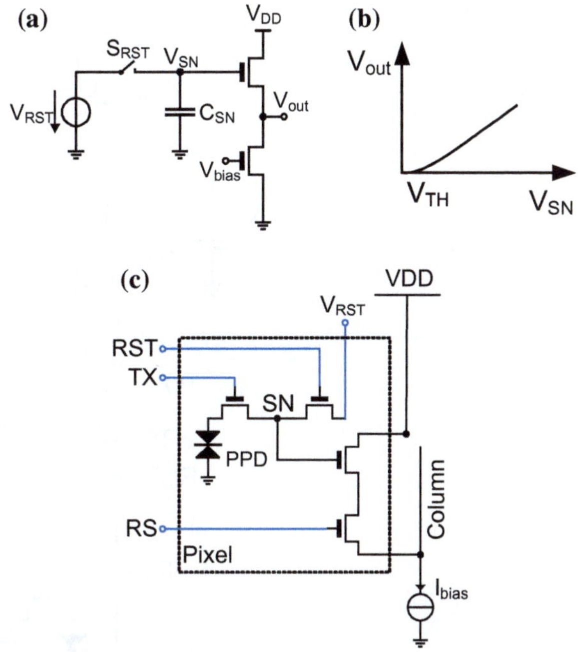

Fig. 2.8 Circuit topology of a source follower stage (**a**), the DC characteristic (**b**) and a schematic of a pixel based on in-pixel source follower buffer (**c**)

The first consists in a large voltage swing. For the 3.3 V transistors usually used in CIS pixels, the voltage swing of the pixel ranges between 1 and 2 V. The second advantage of the source follower scheme is the easy reset. Indeed, the sense node can be reset without much constraints about the exact reset voltage value at the sense node. Hence, this readout scheme is robust against the reset switch imperfections [21]. Figure 2.8(c) shows the schematic of a source follower based 4T pixel. The current bias of the source follower stage is connected to the column. Hence, the current bias is connected to the in-pixel SF transistor only when the row selection switch connects the pixel to the column during the readout.

2.3.2 Common Source Based Pixel

Figure 2.9a shows the schematic of the common source stage with a PMOS load. In this readout scheme, the sense node is connected to the gate of the amplifying transistor and the output voltage is produced at its drain. The output voltage as a function of the sense node voltage is depicted in Fig. 2.9b. This DC characteristic exhibits two main differences compared to the source follower stage. The first consists in the in-pixel voltage gain at the cost of a proportionally lower voltage swing. The second difference consists in the necessity of a precise feedback reset. Indeed, this reset scheme is necessary to make sure that the sense node voltage is at the linear

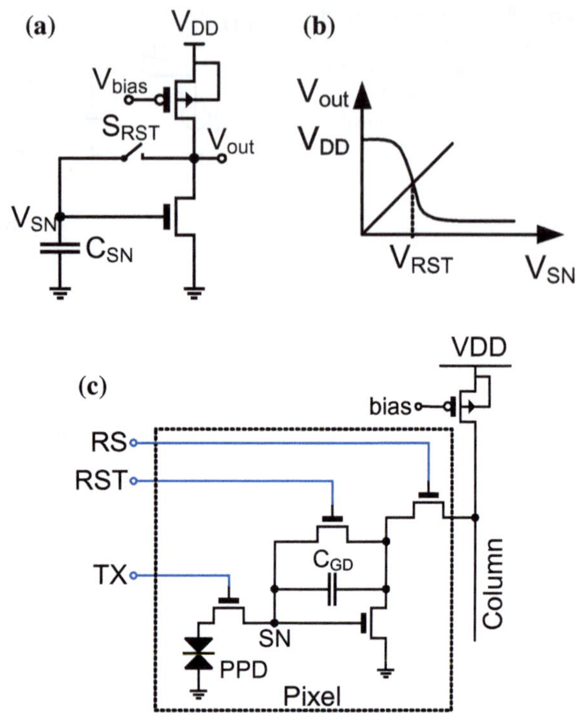

Fig. 2.9 Circuit topology of a common source stage (**a**) the DC characteristic (**b**) and schematic of the pixel based on a common source in-pixel amplification (**c**)

part of the DC characteristic. Note that the common source stage is very sensitive to the reset switch non-ideality. In fact a charge injection induced by this switch can set the sense node voltage in a saturation point.

Figure 2.9(c) shows the schematic of a pixel based on a common source amplification. The in-pixel row selection switch connects the common source transistor the load located at the column level. The load can be implemented by a PMOS, NMOS or a passive resistive element. The common source configuration can be implemented as an open loop gain amplifier [26, 27] or as a capacitive transimpedence amplifier (CTIA) [28, 29] by introducing a capacitive feedback between the common source drain and gate. Note that the open loop gain configuration is a particular case of the CTIA for which the feedback capacitance is simply given by the gate to drain total parasitic capacitance of the in-pixel amplifying transistor.

2.4 Column-Level Amplification

The column-level amplifier is an important block in the readout chain. It is located between the pixel and the next processing stages, namely, the analog buffers, the track-and-holds and analog-to-digital converters. Hence, by introducing the gain

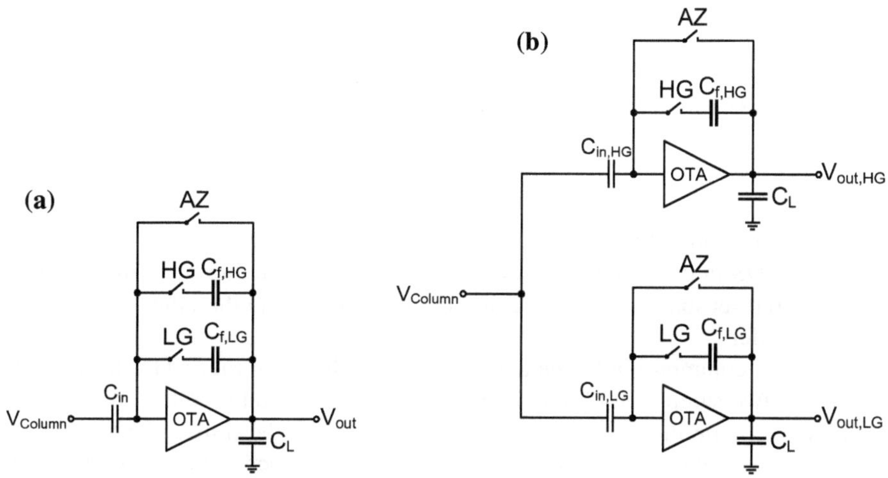

Fig. 2.10 Schematic of the column level switched capacitor amplifier with variable gain (**a**) and dual gain column level amplification

right after the pixel, the noise originating from these stages is minimized. One other important role of the column-level amplifier, especially with a source follower based pixel, is the bandwidth control reducing the thermal noise. The first implementation of column-level amplification leading to a low noise performance was published in [30] reporting a performance of $2\,e^-_{rms}$.

The column-level amplification is generally implemented with switched capacitor amplifiers as shown in Fig. 2.10a. The gain is set by the ratio between the input and feedback capacitors. The amplifier is reset after each readout thanks to an autozero that also dramatically reduces its offset and low frequency noise [21].

In low noise CIS readout chain, the column-level amplifier is especially necessary with the source follower based readout chain. For the best dynamic range, low column level gain is required to exploit as much as possible the full well capacity of the PPD and high column level gain is required for a low noise floor. Hence the column-level amplifier must perform low gain for images under good lighting conditions and high gain in the case of low light imaging. Hence, adaptative column-level gain [31] is required. Figure 2.10b shows the schematic of a parallel dual gain column-level amplifier. In this readout scheme, each pixel is read with both high and low gain levels. In this way, each frame of the imager corresponds to two images, one adequate for dark pixels and the other for highly illuminated pixels. Image post processing can be applied to combine both resulting in a higher dynamic range performance but at the cost of higher power consumption and larger silicon area.

2.5 Correlated Sampling and Analog-to-Digital Conversion

2.5.1 Correlated Double Sampling

Correlated double sampling (CDS) is similar to autozeroing except that the signal is sampled twice and then the difference is taken between these two samples. It has been introduced initially for image processing in CCDs [32, 33] then introduced to CIS. The CDS in the case of CIS is operated by differentiating two samples at the output of the sensor, one after resetting the sense node and the other one after the charge transfer.

The most common implementation of CDS is shown in Fig. 2.11a. In this configuration, two sample-and-hold circuits, connected in parallel to the output of the column level amplifier, are used. Figure 2.11(c) shows the timing diagram of the different control phases operating the readout chain depicted in Fig. 2.11a. First, the in-pixel reset (RST) transistor is switched on in order to reset the sense node. The column level amplifier auto-zero (AZ) switch is closed in order to clear the feedback capacitor C_f, reset the input voltage of the amplifier and store the offset of the amplifier in the input capacitor C_{in}. Once the AZ switch is opened, this offset is canceled. Note that this AZ also reduces the low frequency noise originating from the column level amplifier [21]. The first sample-and-hold circuit switch S_{SH1} is opened after the settling of the voltage corresponding to the reset level. The latter is then held in capacitor C_{SH1}. The TG is then pulsed to its higher level in order to transfer the charge from the PPD to the sense node. The voltage corresponding to the transfer level is then sampled after the signal settles in the second sample-and-hold circuit by opening the second switch S_{SH2}.

Another implementation of the CDS consists in performing the latter at the input of a single slope analog-to-digital converter (ADC). An example of such implementation is shown in Fig. 2.11b. The timing diagram of the control signal of this readout chain is shown in Fig. 2.11(d). During reset of the pixel and the column level amplifier auto-zeroing, the auto-zero switch of the comparator at the input of the ADC stage is closed. When the reset level voltage has settled at the output of the column level amplifier, this switch is opened in order to sample the reset level voltage in the capacitor C_{comp}. Instantaneously, the voltage at the positive input of the comparator becomes equal to the difference between the voltage at the output of the column level amplifier and the reset level voltage sampled in C_{comp}. Now the voltage sampled at the input of the comparator corresponds to the difference between the reset and transfer levels. The ramp then is activated synchronously with the counter and once the rampe voltage is equal to the voltage held at the positive input of the comparator, the latter switches to its high voltage level in order to store the counter value in the SRAM. The CDS time corresponds to the time between the comparator auto-zero and the switching of the comparator output. Note that, in this configuration, the CDS time depends on the signal level.

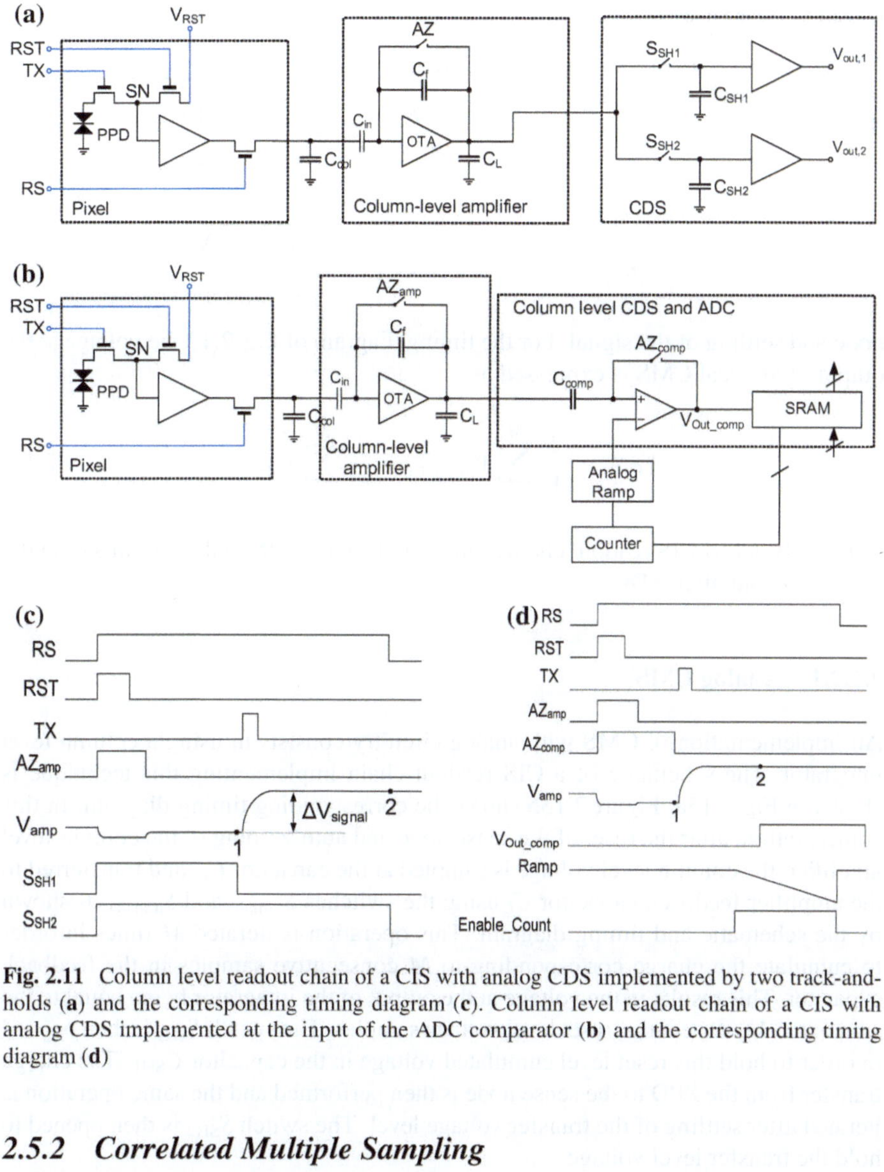

Fig. 2.11 Column level readout chain of a CIS with analog CDS implemented by two track-and-holds (**a**) and the corresponding timing diagram (**c**). Column level readout chain of a CIS with analog CDS implemented at the input of the ADC comparator (**b**) and the corresponding timing diagram (**d**)

2.5.2 Correlated Multiple Sampling

Correlated multiple sampling (CMS) was introduced for CIS in [34–36]. It combines CDS with averaging. CMS of order M corresponds to averaging M samples at the reset level and M samples after charge transfer and then differentiating the two averages. Figure 2.12 shows the timing diagram of CMS in case for a CIS readout chain. A CMS of order M corresponds to averaging M samples in of the reset level voltage and M other samples after transferring the charge from the PPD to the sense

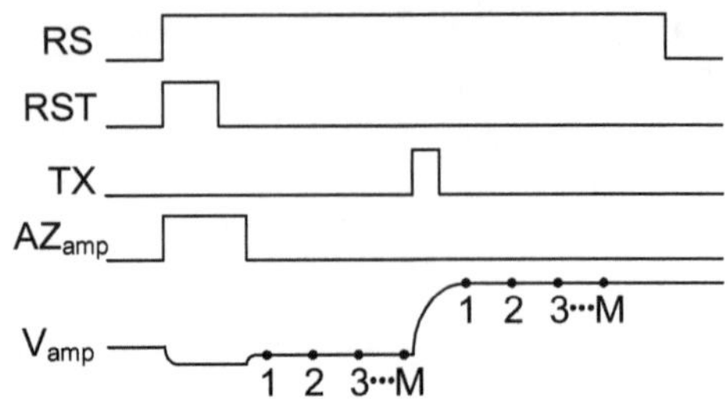

Fig. 2.12 Timing diagram of the an M-order CMS showing the M samples at the first and second voltage levels

node and settling of the signal. For the timing diagram of Fig. 2.12 the voltage at the output of an ideal CMS is expressed as:

$$V_{CMS} = \frac{1}{M}\sum_{i=1}^{M} V_{\text{transfer},i} - \frac{1}{M}\sum_{i=1}^{M} V_{\text{reset},i}. \tag{2.19}$$

Practically, CMS CIS readout chains can be performed with analog circuits or in the digital domain after ADC.

2.5.2.1 Analog CMS

An implementation of CMS with analog circuitry consists in using a column level integrator. The schematic of a CIS readout chain implementing this technique is shown in Fig. 2.13a. Figure 2.13b shows the corresponding timing diagram. In this configuration, after the reset of the sense node and auto-zeroing of the column-level amplifier, the column level voltage is sampled at the capacitor C_{in} and transferred to the amplifier feedback capacitor C_{f} using the switches S_{sample} and S_{transfer} as shown by the schematic and timing diagram. This operation is iterated M times in order to cumulate the charge corresponding to M consecutive samples in the feedback capacitor. This results in the voltage at the output of the integrator being equal to the sum of the M consecutive reset level voltage samples. The switch S_{SH1} is then opened in order to hold this reset level cumulated voltage in the capacitor C_{SH1}. The charge transfer from the PPD to the sense node is then performed and the same operation is iterated after settling of the transfer voltage level. The switch S_{SH2} is then opened to hold the transfer level voltage.

The integrator results hence in cumulating the reset level and transfer level samples instead of averaging. Thus, this configuration results in a dynamic range decrease by a factor M. An alternative to this implementation [36] consists in using folding integration by introducing a digitally assisted feedback that prevents the integrator from saturating. This implementation is obtained at the cost of additional active mixed signal circuitry.

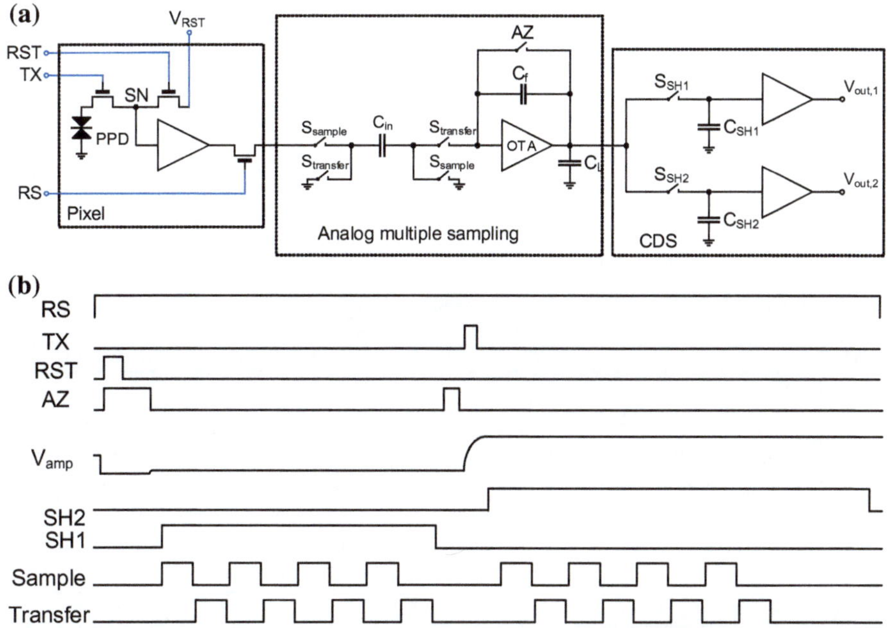

Fig. 2.13 Analog implementation of CMS using column-level integrator (**a**) and the corresponding timing diagram (**b**)

2.5.2.2 Digital CMS

CMS can also be performed in the digital domain [37, 38]. Figure 2.14b shows a readout chain based on a 4 T pixel, column-level amplification and multiple rampe ADC used to perform CMS by multiple analog-to-digital conversions using multiple ramping. The corresponding timing diagram is shown in Fig. 2.14a. After the sense node reset, the column level amplifier auto-zeroing and settling of the reset level voltage at the input node of the ADC comparator. Multiple ramps are iterated synchronously with a bitwise counter. The counts corresponding to M conversions are cumulated. Then the charge is transferred from the PPD to the sense node and the bitwise inversion is activated in order to set the counter in the count down mode. Similarly to the reset level voltage. The multiple ramping is activated in the same way after settling the transfer voltage level at the output of the column-level amplifier. But in this case the counter counts down subtracting in this way the cumulated transfer level voltage samples from the reset level ones.

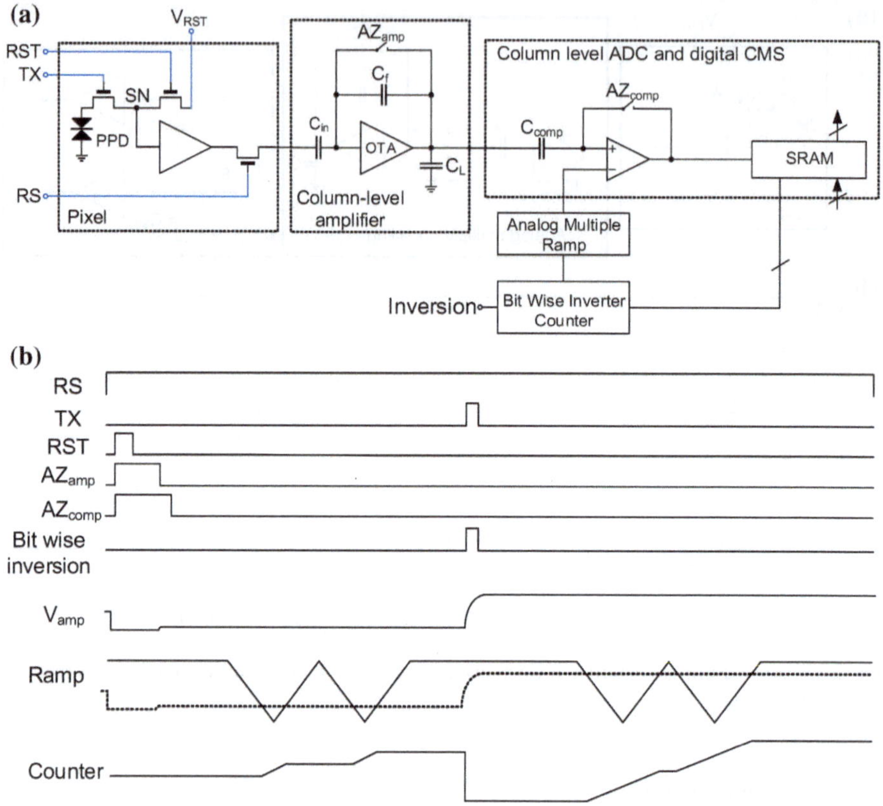

Fig. 2.14 Digital implementation of CMS using column-level multiple ramp ADC (**a**) and the corresponding timing diagram (**b**)

2.6 Summary

PPDs are the key element in low noise CIS. PPDs present a buried potential well in a depleted area that collects the photoelectrons. Low noise CIS pixels comprise, in addition to the PPD, at least four transistors to control the selection, reset and transfer. The fourth transistor is generally operated in the source follower configuration to buffer the voltage level of the sense node. The latter can also be used in the common source configuration for a higher pixel-level voltage gain. The pixel readout is performed in a double sampling scheme and the signal corresponds to the difference between the sense node voltage right after the reset and its level after the charge transfer from the PPD. This readout scheme cancels the reset sampled noise and reduces the low frequency noise originating from the readout chain electronics. CIS chips are generally built in a column parallel structure. The pixels of each line are simultaneously selected and read. The column level circuits include the bias current or load of the in-pixel amplification stage, the column-level amplifiers are used to introduce a gain minimizing the noise contribution of the next stages and controlling the bandwidth for optimal thermal noise. The CDS and ADC take place after column amplification and can be implemented in different schemes. Signal multiple sampling and averaging can also be implemented for further noise reduction. In the low noise CIS context, the combination of averaging and double sampling is called CMS. It can be implemented with analog or digital circuitry.

References

1. G.P. Weckler, A silicon photodevice to operate in a photon flux integrated mode, in *Electron Devices Meeting, 1965 International*, vol. 11 (1965), pp. 38–39
2. G.P. Weckler, Operation of p-n junction photodetectors in a photon flux integrating mode. IEEE J. Solid-State Circuits **2**(3), 65–73 (1967)
3. P.J.W. Noble, Self-scanned silicon image detector arrays. IEEE Trans. Electron Devices **15**(4), 202–209 (1968)
4. L. Walsh, R. Dyck, A new charge-coupled area imaging device, in *Proceedings of CCD Applicational Conference*, September 1973, pp. 21–22
5. E.R. Fossum, D.B. Hondongwa, A review of the pinned photodiode for CCD and CMOS image sensors. IEEE J. Electron Devices Soc. **2**(3), 33–43 (2014)
6. N. Teranishi, A. Kohono, Y. Ishihara, E. Oda, K. Arai, No image lag photodiode structure in the interline CCD image sensor, in *Electron Devices Meeting, 1982 International*, vol. 28 (1982), pp. 324–327
7. B.C. Burkey, W.C. Chang, J. Littlehale, T.H. Lee, T.J. Tredwell, J.P. Lavine, E.A. Trabka, The pinned photodiode for an interline-transfer CCD image sensor, in *Electron Devices Meeting, 1984 International*, vol. 30 (1984), pp. 28–31
8. E.R. Fossum, Active pixel sensors: are CCDs dinosaurs? in *Proceedings of SPIE*, vol. 1900 (1993), pp. 2–14, http://dx.doi.org/10.1117/12.148585
9. E.R. Fossum, Ultra-low-power imaging systems using CMOS image sensor technology, in *Proceedings of SPIE*, vol. 2267 (1994), pp. 107–111, http://dx.doi.org/10.1117/12.187470
10. P.P.K. Lee, R.C. Gee, R.M. Guidash, T.-H. Lee, E.R. Fossum, An active pixel sensor fabricated using CMOS/CCD process technology, in *Proceedings of IEEE Workshop on CCDs and Advanced Image Sensors*, April 1995
11. R.M. Guidash, T.H. Lee, P.P.K. Lee, D.H. Sackett, C.I. Drowley, M.S. Swenson, L. Arbaugh, R. Hollstein, F. Shapiro, S. Domer, A 0.6 /spl mu/m CMOS pinned photodiode color imager technology, in *Electron Devices Meeting, IEDM '97, Technical Digest, International* (1997), pp. 927–929
12. I. Inoue, N. Tanaka, H. Yamashita, T. Yamaguchi, H. Ishiwata, H. Ihara, Low-leakage-current and low-operating-voltage buried photodiode for a CMOS imager. IEEE Trans. Electron Devices **50**(1), 43–47 (2003)
13. K. Yonemoto, H. Sumi, R. Suzuki, and T. Ueno, A CMOS image sensor with a simple fpn-reduction technology and a hole accumulated diode, in *Solid-State Circuits Conference, 2000. Digest of Technical Papers, ISSCC 2000 IEEE International*, February 2000, pp. 102–103
14. S.R. Shortes, W.W. Chan, W.C. Rhines, J.B. Barton, D.R. Collins, Development of a thinned, backside-illuminated charge-coupled device imager, in *Electron Devices Meeting, 1973 International*, vol. 19 (1973), pp. 415–415
15. S. Shortes, W. Chan, W. Rhines, J. Barton, D. Collins, Characteristics of thinned backside-illuminated charge-coupled device imagers. Appl. Phys. Lett. **24**(11), 565–567 (1974)
16. C. Chen, Z. Bing, W. Longsheng, L. Xin, W. Junfeng, Pinch-off voltage modeling for CMOS image pixels with a pinned photodiode structure. J. Semiconductors **35**(7), 074012 (2014), http://stacks.iop.org/1674-4926/35/i=7/a=074012

17. Z. Huiming, W. Tingcun, Z. Ran, Modeling of pinned photodiode for CMOS image sensor, in *2011 IEEE International Conference on Signal Processing, Communications and Computing (ICSPCC)*, September 2011, pp. 1–4
18. A. Pelamatti, V. Goiffon, M. Estribeau, P. Cervantes, P. Magnan, Estimation and modeling of the full well capacity in pinned photodiode cmos image sensors. IEEE Electron Device Lett. **34**(7), 900–902 (2013)
19. S. Sze, K. Ng, *Physics of Semiconductor Devices* (Wiley, New York, 2006), pp. 79–134, https://books.google.ch/books?id=o4unkmHBHb8C
20. S. Park, H. Uh, The effect of size on photodiode pinch-off voltage for small pixel CMOS image sensors. Microelectron. J. **40**(1), 137–140 (2009)
21. C. Enz, G. Temes, Circuit techniques for reducing the effects of op-amp imperfections: autozeroing, correlated double sampling, and chopper stabilization. Proc. IEEE **84**(11), 1584–1614 (1996)
22. J.R. Janesick, Scientific charge-coupled devices, in *SPIE* (2001), pp. 3–93
23. S.G. Chamberlain, Photosensitivity and scanning of silicon image detector arrays. IEEE J. Solid-State Circuits **4**(6), 333–342 (1969)
24. S.G. Chamberlain, V.K. Aggarwal, Photosensitivity and characterization of a solid-state integrating photodetector. IEEE J. Solid-State Circuits **7**(2), 202–204 (1972)
25. B. Razavi, *Design of Analog CMOS Integrated Circuits* (McGraw-Hill Publications, New York, 2001), pp. 47–92
26. C. Lotto, P. Seitz, T. Baechler, A sub-electron readout noise CMOS image sensor with pixel-level open-loop voltage amplification, in *2011 IEEE International Solid-State Circuits Conference*, February 2011, pp. 402–404
27. J. Yang, K.G. Fife, L. Brooks, C.G. Sodini, A. Betts, P. Mudunuru, H.-S. Lee, A 3Mpixel low-noise flexible architecture CMOS image sensor, in *2006 IEEE International Solid-State Circuits Conference, ISSCC 2006, Digest of Technical Papers*, February 2006, pp. 2004–2013
28. B. Fowler, Single photon CMOS imaging through noise minimization, in *Single-Photon Imaging* (Springer, Heidelberg, 2011), pp. 159–194
29. K. Murari, R. Etienne-Cummings, N.V. Thakor, G. Cauwenberghs, A CMOS in-pixel ctia high-sensitivity fluorescence imager. IEEE Trans. Biomed. Circuits Syst. **5**(5), 449–458 (2011)
30. A. Krymski, N. Khaliullin, H. Rhodes, A 2 e- noise 1.3 megapixel CMOS sensor, in *Proceedings of IEEE Workshop on CCDs and Advanced Image Sensors*, May 2003
31. M. Sakakibara, S. Kawahito, D. Handoko, N. Nakamura, H. Satoh, M. Higashi, K. Mabuchi, H. Sumi, A high-sensitivity CMOS image sensor with gain-adaptive column amplifiers. IEEE J. Solid-State Circuits **40**(5), 1147–1156 (2005)
32. M.H. White, D.R. Lampe, F.C. Blaha, I.A. Mack, Characterization of surface channel CCD image arrays at low light levels. IEEE J. Solid-State Circuits **9**(1), 1–12 (1974)
33. R.J. Kansy, Response of a correlated double sampling circuit to 1/f noise [generated in CCD arrays]. IEEE J. Solid-State Circuits **15**(3), 373–375 (1980)
34. N. Kawai, S. Kawahito, Effectiveness of a correlated multiple sampling differential averager for reducing 1/f noise. IEICE Electronics Express **2**(13), 379–383 (2005)
35. Y. Chen, Y. Xu, A.J. Mierop, A.J.P. Theuwissen, Column-parallel digital correlated multiple sampling for low-noise CMOS image sensors. IEEE Sens. J. **12**(4), 793–799 (2012)
36. S. Suh, S. Itoh, S. Aoyama, S. Kawahito, Column-parallel correlated multiple sampling circuits for CMOS image sensors and their noise reduction effects. Sensors **10**(10), 9139–9154 (2010)
37. Y. Lim, K. Koh, K. Kim, H. Yang, J. Kim, Y. Jeong, S. Lee, H. Lee, S.-H. Lim, Y. Han, J. Kim, J. Yun, S. Ham, Y.-T. Lee, A 1.1 e- temporal noise 1/3.2-inch 8 mpixel CMOS image sensor using pseudo-multiple sampling, in *2010 IEEE International Solid-State Circuits Conference Digest of Technical Papers (ISSCC)* (2010), pp. 396–397
38. Y. Chen, Y. Xu, Y. Chae, A. Mierop, X. Wang, A. Theuwissen, A 0.7e-rms temporal-readout-noise CMOS image sensor for low-light-level imaging, in *2012 IEEE International Solid-State Circuits Conference Digest of Technical Papers (ISSCC)*, February 2012, pp. 384 –386

Chapter 3
Noise Sources and Mechanisms in CIS

The operation principles of the conventional low noise CIS presented in the previous Chapter can suffer from non-idealities, defects and random fluctuations at different levels corrupting the integrity of the signal. These random events occur at the level of the PPD, during the charge transfer and also at the level of the readout circuit electronics. In this chapter, the noise sources corrupting the signal in CIS from the front to the back-end are reviewed and illustrated with experimental examples.

3.1 Photon Shot Noise

The shot noise is a statistical phenomenon appearing in nature for physical processes resulting from a series of independent events occurring with the same probability. Consider the case of a particle flux with a rate λ constant in time. In the time interval $[0, t]$ the average number of incident particles is given by $\lambda \cdot t$. If each particle has the same probability of incidence p, then the number of incident particles obeys to a Binomial Law. If the probability p is close to unity, this Binomial law tends to a Poisson distribution where the probability to receive n particles in the interval $[0, t]$ is given by [1]:

$$p_n(t) = \frac{(\lambda \cdot t)^n}{n!} \cdot e^{-\lambda \cdot t}. \tag{3.1}$$

The property characterizing the shot noise is the fact that the variance of the number of received particles is equal to the constant average number. In optics and electronics, the quantized nature of light and charge makes a photon or electron flux obey this Poisson process. The photon incidence obeys so accurately to the Poisson distribution that it can be used to characterize photosensors [2] and particularly CIS. Consider a CIS with a conversion gain A_{CG}. If the imager receives a constant photon flux

A. Boukhayma, *Ultra Low Noise CMOS Image Sensors*,
Springer Theses, https://doi.org/10.1007/978-3-319-68774-2_3

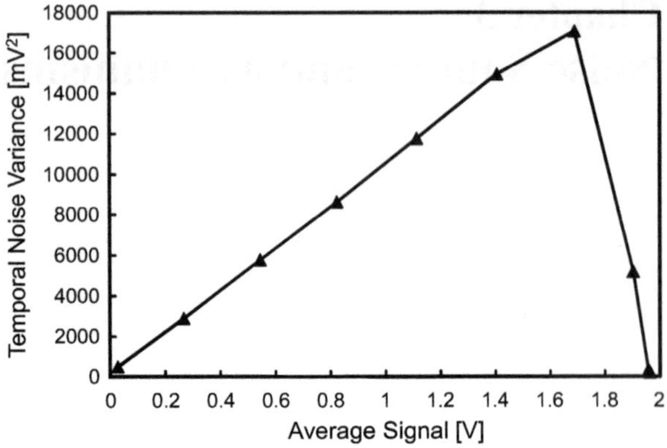

Fig. 3.1 PTC of a CIS From the test chip designed and fabricated in the frame of this work verifying (3.2)

corresponding to N photons for an integration time T_{int}, then the variance of the output signal can be expressed as

$$Var[V_{\mathrm{out}}(N)] = A_{\mathrm{CG}} \cdot E[V_{\mathrm{out}}(N)] + Var[V_{\mathrm{out}}(0)], \tag{3.2}$$

where $Var[V_{\mathrm{out}}(0)]$ denotes the output signal variance in the dark which is the read noise of the imager. The variance versus average output signal curve is called the photon transfer curve (PTC) [2]. From this curve, the conversion gain A_{CG} is measured as the slope of its linear part. The average signal for which the PTC curve collapses corresponds to the full well capacity of the imager. This saturation point can be due to the saturation of the electronic readout circuit or to the saturation of the photodiode in case the dynamic range of the readout chain is high enough. Figure 3.1 shows the PTC obtained with the image sensor designed in the frame of this work. The curve demonstrates that the shot noise dominates in the linear part of the curve as expected by (3.2). More details about this measurement will be given in Chaps. 6 and 7.

3.2 Dark Current

The dark current is the name commonly given to the charge generation rate at the photosensor in the absence of light. This phenomenon is intrinsic to semiconductor devices and presents the first element that can corrupt the signal in the readout chain. The dark current is the parameter limiting the integration time of any solid state image sensor and the only way to overcome this limitation, for a given device, is cooling [3]. The physical process behind the dark current is the different generation recombination mechanisms occurring in the depleted area of the PPD and directly accumulated in the potential well or outside the depleted region and diffusing toward it.

The trap assisted generation recombination is the most common physical model for dark charge generation. Based on this physical model, the impurities introduced to the semiconductor may be associated with energy states between the valence and conductance band in the range where no energy states are available for carriers in intrinsic semiconductor. Hence these impurities act as steps facilitating the transition of electrons between the valance and conductance band as shown in Fig. 3.2. This process is also known as the "hopping conduction".

Figure 3.3 shows different dark current sources in a CIS PPD. Dark current can be generated at the level of the depleted area of the PPD. It can also be generated at the level of the bulk in the field free area and then diffuse to the potential well. It is known that the trap density increases at the surface and interfaces between different materials due to impurities and process defects. Hence it is believed that an important part of the dark current generation occurs at the level of the PPD surface and the $Si\text{-}SiO_2$ interface under the transfer gate and at the level of interfaces with shallow trench channel (STI).

The dark current generation in semiconductors is governed by the Shockley-Hall-Read equation [4–6] where the net carrier generation/recombination rate U_{SHR} (carriers $\cdot\, s^{-1} \cdot cm^{-3}$) is expressed as:

$$U_{\mathrm{SHR}} = \frac{\sigma_{\mathrm{p}}\sigma_{\mathrm{n}}U_T(pn - n_{\mathrm{i}}^2)N_{\mathrm{t}}}{\sigma_{\mathrm{n}}\left(n + n_{\mathrm{i}}exp(\frac{E_{\mathrm{t}}-E_{\mathrm{Fi}}}{kT})\right) + \sigma_{\mathrm{p}}\left(p + n_{\mathrm{i}}exp(\frac{E_{\mathrm{Fi}}-E_{\mathrm{t}}}{kT})\right)}, \tag{3.3}$$

where σ_{n} and σ_{p} refer to the electron and hole cross sections (cm^2), U_T is the thermal voltage, p and n are the electron and hole concentrations (cm^{-3}), n_{i} is the intrinsic carrier concentration (electon-hole/cm^3). $E_{\mathrm{t}} - E_{\mathrm{Fi}}$ is the gap between the energy of the trap and the intrinsic Fermi level, k is the Boltzmann constant and T is the absolute temperature.

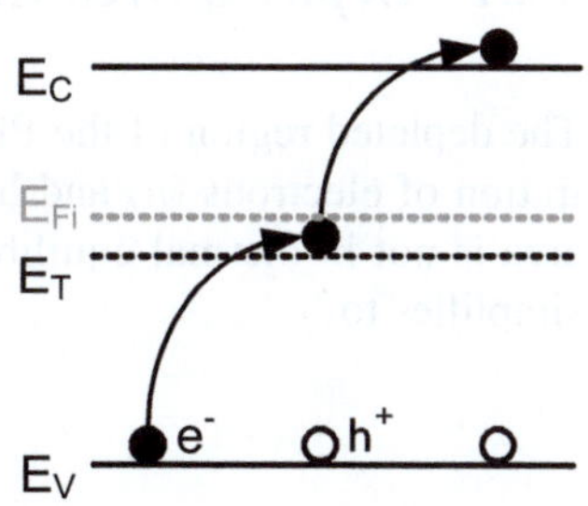

Fig. 3.2 Trap assisted carrier generation "hopping conduction"

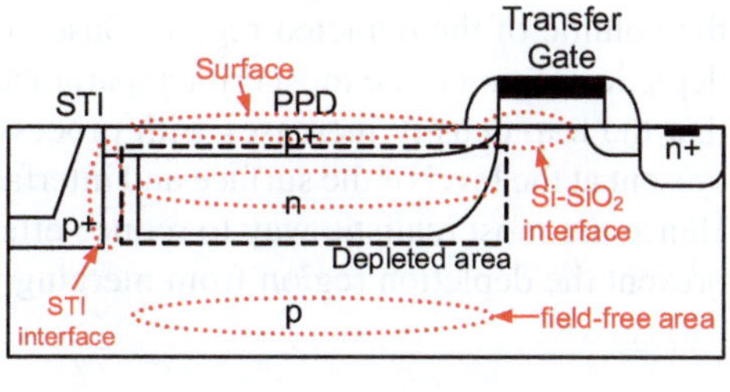

Fig. 3.3 Dark current carriers generation sources

In thermal equilibrium, the product of electrons and holes concentration verifies [7]:

$$n \cdot p = n_i^2 \tag{3.4}$$

In doped semiconductors, the density of majority carriers is rather given by the doping concentration. Hence the holes concentration in an n-doped semiconductor is given by

$$p = \frac{n_i^2}{N_D}, \tag{3.5}$$

where N_D is the doping concentration of donors. Similarly, the electrons concentration in a p-doped semiconductor is given by

$$p = \frac{n_i^2}{N_A}, \tag{3.6}$$

where N_A is the density of acceptors.

Equation (3.3) is the starting point to analyze the generation recombination processes for the PPD. Note that, in thermal equilibrium, $n \cdot p$ is equal to n_i^2, resulting in a zero generation/recombination rate and absence of dark current. But the doped semiconductor is not in thermal equilibrium during depletion. Hence the depleted regions in the 4T pixel are critical in the dark current generation. In case of non thermal equilibrium, the traps with energies E_t close to the intrinsic Fermi level energy E_{Fi} of the semiconductor are the major contributors to the dark current generation.

3.2.1 Depleted Area Generation

The depleted region of the PPD is deserted from its majority carriers. The concentration of electrons (n) and holes (p) are negligible compared to n_i. Note that this area is not in thermal equilibrium, hence (3.4) is not verified. Consequently, (3.3) simplifies to

$$U_{dep} = \frac{-\sigma U_T n_i^2 N_t}{2cosh(\frac{E_t - E_{Fi}}{kT})}. \tag{3.7}$$

The total dark current is given by the integral of (3.7) over the trap energies and over the volume of the depleted region. Based on (3.7), the density of traps N_t over the depletion region is the most critical parameter in dark current generation. It is known that the trap density increases with process defects and impurities which are more present at the level of the surface and interfaces. These process defects are inevitable. Hence the most realistic way to reduce effectively this dark current generation is to prevent the depletion region from merging with the surface and interfaces.

3.2.1.1 Surface and STI Interface

At the level of the PPD surface, the p+ layer on top of the PPD n-well plays a crucial role in the dark current reduction. Indeed thanks to this layer, the vertical potential profile of the PPD pushes the positive charge carriers (majority carriers in the p+ layer) towards the surface preventing in this way the depletion region of the PPD from merging with the surface states. Moreover the electrons that may be generated by the surface states would recombine before reaching the depleted area thanks to the high density of holes at the level of the surface.

The same principle described above is used to reduce the dark current generated at the level of the interface with the STI oxide interface. Indeed, a p+ layer is introduced at the level of the STI oxide interface as shown in Fig. 3.3.

3.2.1.2 Transfer Gate Oxide Interface

The states located at the Si-SiO_2 interface below the transfer gate are a major contributor to the dark current generation. With tiny transfer gates, this dark current generation may behave as random telegraph signal (RTS) [8] causing a minority of "blinking" pixels. This dark current contribution increases when the depleted area of the PPD merges with the depleted area under the transfer gate. Another mechanism generating the dark current at the level of the transfer gate oxide interface is the lateral field created by the overlap between the PPD p+ layer and the transfer gate [9]. It has been reported in [9] that this strong field increases the dark current by causing impact ionisation creating more traps. The most common way to reduce this dark current consists in driving the transfer gate with a negative voltage during the integration time. Indeed a negative voltage at the level of the transfer gate suppresses the depletion under the transfer gate so that the latter does not merge with the PPD depletion region. Furthermore, an accumulation layer of positive carriers is created under the transfer gate thanks to the negative voltage [8, 10, 11]. These accumulated holes play the same role as the p+ layer on the surface of the PPD by recombining with the charges generated by the oxide traps preventing them from being cumulated in the PPD potential well.

3.2.2 *Field-Free Area Generation*

Unlike the depleted region of the PPD, The deep field-free area under the potential well remains in thermal equilibrium. Hence, the generation/recombination dark current is negligible as discussed above. Another dark current mechanism takes place in this deep area which is the diffusion dark current. The concentration of electrons in the deep epitaxial p region under the PPD is given by $\frac{n_i^2}{N_A}$. However this concentration decreases exponentially in the direction of the depleted area of the PPD. It is expressed as [3]

$$n = \frac{n_i^2}{N_A}(1 - e^{-\frac{x}{L_n}}), \tag{3.8}$$

where x is the distance from the depletion edge and L_n denotes the diffusion length for electrons. This electron concentration gradient gives rise to a diffusion current. Indeed the electrons would move from the high concentration to the low concentration region (depleted area). This diffusion current can be expressed as:

$$I_{diff} = qD_n\frac{dn}{dx} = \frac{qD_n n_i^2}{L_n N_A}, \tag{3.9}$$

where D_n is the diffusion coefficient ($cm^2 \cdot s^{-1}$) expressed by the Einstein relation as $D_n = \mu_n U_T$ with μ_n being the electron mobility in silicon.

The diffusion dark current is inversely proportional to the diffusion length which is an indicator of the semiconductor quality. Indeed the diffusion length is greater in semiconductors with a lower density of traps [3]. Consequently, the field free area dark current depends on the manufacture and represents a foundry quality indicator.

3.2.3 *Dark Current Shot Noise*

The generation of charge carriers in the dark results in a steady current that it generally measured for image sensors as number of electrons per second or as a current density per unit area of the PPD device. Typical values in state-of-the-art imagers are in the order of a few $e^- \cdot s^{-1} \cdot \mu m^{-2}$ at room temperature. The noise caused by the dark current manifests itself by a shot noise associated to this steady current. The dark current shot noise variance, after integration, is hence given by:

$$\overline{Q_{dark}^2} = \cdot I_{dark} \cdot t_{int}, \tag{3.10}$$

where I_{dark} si the dark current in $e^- \cdot s^{-1}$ and t_{int} is the integration time of the imager.

3.3 Transfer Noise

3.3.1 *Charge Transfer Non-idealities*

The charge transfer from the PPD to the SN can be affected by the noise related to the charge deficit due to incomplete transfer and lag [12]. Unlike the sampled reset kTC noise, this noise is not canceled by the correlated double sampling. The charge transfer noise has been extensively studied for CCDs [13, 14] because an efficient charge transfer is crucial in such devices. For 4T pixels based on PPDs, this issue is becoming an active point of research [9, 12, 15] with the read noise reaching deep subelectron levels in standard CIS process [16, 17]. The main imperfections affecting the signal during the charge transfer from the PPD potential well to the sense node are shown in Fig. 3.4. Process level techniques are used to introduce a fringing field in order to shape the potential well in a way that drifts the electrons toward the transfer gate. The aim of this process level optimization is to accelerate the transfer. This fringing field may not be applied ideally, in which case the electrons cumulated in the potential well would rather diffuse toward the transfer gate before being drifted to the sense node. This non ideality results hence in a slow transfer reducing the charge transfer efficiency. The potential shape under the transfer gate my also suffer from some imperfections that can take the form of barriers or charge pockets. These defects may prevent or slow down part of the photoelectrons during the transfer. Once the transfer gate potential barrier is raised, part of these charges goes back to the PPD instead of ending in the sense node. Finally, if the PPD cumulates a large number of photoelectrons or if the reset voltage of the sense node is too low, the charges are not completely transferred to the sense node and a part remains under the transfer gate. When the transfer gate barrier is raised, part of this charge goes back to the PPD. This process is referred to as the "spill back". All the nonidealities reported below result in a part of the total photoelectrons being left in the potential well of the PPD after each readout. This process is quantified by the charge transfer inefficiency (CTI) calculated as the average percentage of the charge left after each readout. The noise related to this left charge behaves as a shot noise [12] similarly to buried channel CCDs [14]. The noise charge variance is hence given by:

$$\overline{Q_{\mathrm{CTI}}^2} = CTI \cdot N, \tag{3.11}$$

where N denotes the number of photoelectrons. CTI can be measured by performing two consecutive readouts after a relatively large integration time. This operation is iterated several times and the measured outputs of the first and second readouts are averaged. The ratio of the two averaged values determines the value of the CTI. In state-of-the-art CIS with 4T pixels, values of lag as low as 0.1% have been reported. Measurements presented later in Chap. 7 confirm this result for a CIS designed in a 180 nm process. Thus the shot noise associated to the lag can be neglected compared to the read noise in the low light context.

Another imperfection introducing noise during the charge transfer is the presence of traps at the level of the Si-SiO_2 interface under the transfer gate. A large positive voltage is applied to the transfer gate in order to create a voltage difference accelerating the diffusion of the cumulated charges to the sense node. But this positive voltage attracts the electrons to the Si-SiO_2 interface since there is no built-in potential under the transfer gate. Hence some electrons may be trapped during the transfer process. Unlike the noise introduced by the lag, the variance of this fluctuation is not easy to measure. Indeed, trapping can be, for instance, confused with the impact of a

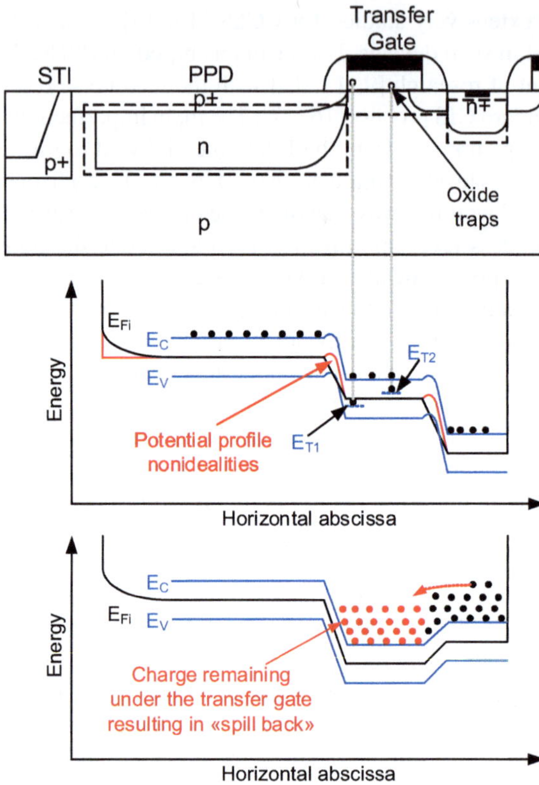

Fig. 3.4 Nonidealities of the charge transfer process

charge pocket. It has been shown in [15] that these events can be analyzed separately by increasing the transfer time. In fact, the non-idealities related to the potential shape slow down the charge carriers transfer, but for a transfer time large enough, the diffusion takes over and leads the electrons to the sense node deep potential well.

3.4 Electronic Noise

This Section covers the uncorrelated noise sources corrupting the signal at the level of the readout circuitry. These include the thermal noise characterized by fast fluctuations, the 1/*f* noise characterized by slow fluctuations and the random telegraph signal noise (RTS) characterized by random switching events between two or more states. Figure 3.5 presents three cases where one of these sources dominates the noise in the measured signal. The shot noise associated to the leakage current is also discussed. The origins of these noise sources are reviewed and the derivation of their power spectral density (PSD) is presented.

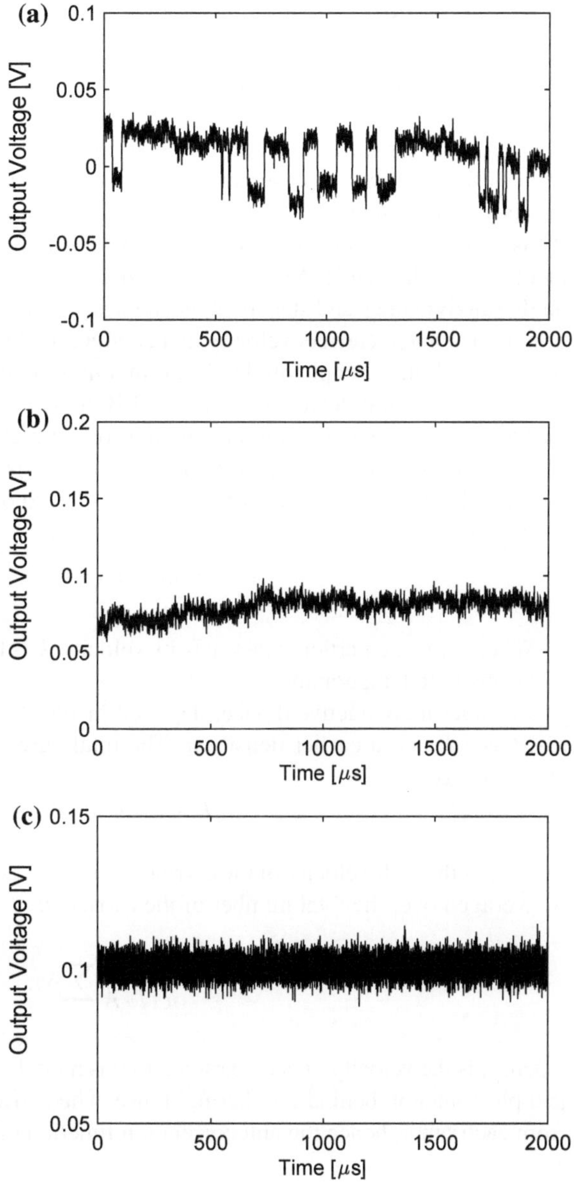

Fig. 3.5 Measured voltage, with a gain of 64, at the output of the in-pixel source follower transistors pixels (From the test chip designed and fabricated in the frame of this work): **a** features a dominant RTS noise in addition to thermal noise, **b** features only 1/*f* and thermal noise (**c**) features a dominant thermal noise

3.4.1 Thermal Noise

3.4.1.1 Thermal Noise in Resistors

Thermal noise is a fundamental physical phenomenon observed in all conducting devices at a positive absolute temperature. Like the shot noise, thermal noise behavior obeys to the same statistical law independently of the conducting device material. It has been first measured by J.B. Johnson in 1928 at Bell Labs. Based on his measurements, his colleague H. Nyquist proposed a first derivation of thermal noise based on thermodynamics and statistical mechanics. The origin of this noise is the fluctuation of the charge carriers velocity in a conducting device due to thermal excitation. A simplified microscopic model of thermal noise consists in assimilating the carriers in a conducting device to a Maxwell-Boltzmann idealized gaz where particles exchange energy only by mutual collisions or thermally. The random motion resulting from these collisions, called Brownian motion, is governed by the equipartition theorem that relates the mean kinetic energy of each particle (carrier) to the ambiant temperature as

$$\frac{1}{2}m\overline{v^2} = \frac{1}{2}kT, \tag{3.12}$$

Where m is the particle mass, v is its velocity, k is the Boltzmann constant and T is the absolute temperature.

Consider a conductive device, depicted in Fig. 3.6, with a gap L, a section A, a resistance R and a carrier density n. The total current crossing this device can be expressed as

$$I = q \cdot A \cdot n \cdot v. \tag{3.13}$$

where v is the drift velocity of the carriers, along the axis perpendicular to the section A, averaged over the total number of the carriers $n \cdot l \cdot A$. It can be expressed as

$$v = \frac{1}{n \cdot l \cdot A}\sum_i v_i, \tag{3.14}$$

where v_i is the velocity of each particle. In this model, the fluctuation of v is the physical phenomenon behind the thermal noise. The carriers velocities are independent from each other, hence the autocorrelation function of v can be expressed as

$$R_v(t) = \frac{1}{(n \cdot l \cdot A)^2}\sum_i R_{v_i}(t). \tag{3.15}$$

Each carrier conserves the same velocity until it collides with another carrier. The velocities before and after collision are uncorrelated. Let τ_c be the relaxation time

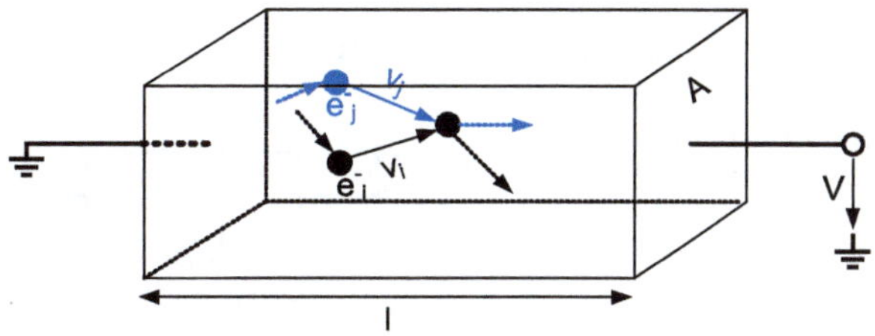

Fig. 3.6 Microscopic physical model of thermal noise in resistor of section A and gap l

related to the collisions. The probability that the carrier does not collide with another one during t is given by $e^{-\frac{t}{\tau_c}}$. Hence, the autocorrelation function can be expressed as

$$R_{v_i}(t) = \overline{v^2} \cdot e^{-\frac{t}{\tau_c}}. \tag{3.16}$$

The current autocorrelation function can now be expressed using (3.12), (3.15) and (3.16). The current noise PSD is given by its Fourier transform as

$$S_I(f) = 2 \cdot k \cdot T \frac{q^2 \cdot n \cdot A \cdot \tau_c}{l \cdot m} \cdot \frac{1}{1 + (2\pi f \tau_c)^2}. \tag{3.17}$$

The resistance is defined by $R = \frac{l}{\sigma \cdot A}$, where, in the device physics, the conductivity σ is given by $\sigma = q \cdot n \cdot \mu$ and the mobility μ is defined by $\mu = q\frac{\tau_c}{m}$. Hence, the unilateral expression of the current thermal noise PSD crossing the resistance R is expressed as

$$S_I(f) = \frac{4kT}{R} \cdot \frac{1}{1 + (2\pi f \tau_c)^2}. \tag{3.18}$$

Note that $f\tau_c \ll 1$ for conventional electronic circuits and the expression simplifies to $\frac{4kT}{R}$. The corresponding voltage fluctuation has a PSD given by:

$$S_V(f) \simeq 4kTR. \tag{3.19}$$

The thermal noise of a resistor is modeled at the circuit level by a noise current source in parallel to the ideal resistor with a current noise PSD of $\frac{4kT}{R}$. It can be also modeled by a voltage source with a noise PSD of $4kTR$ in series with the ideal resistor as shown in Fig. 3.7a.

3.4.1.2 KT/C Noise

kT/C noise is the name commonly given to the fluctuation of the voltage across a capacitor as a result of the thermal noise originating from a resistive element connected to one of its terminals as shown in Fig. 3.7b. Consider a capacitor C connected in series to a resistor R. The thermal noise originating from the resistor

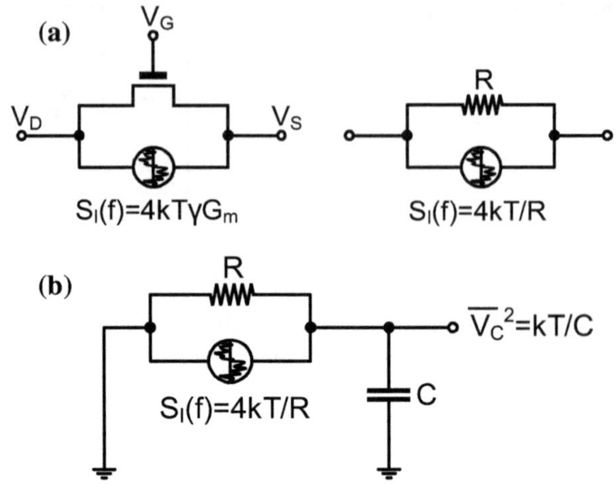

Fig. 3.7 Thermal noise source model for a MOS transistor and a resistor (**a**) and kT/C noise mechanism (**b**)

can be modeled by a voltage source with a PSD $4kTR$ low pass filtered by the RC linear circuit. The noise variance of the voltage across the capacitor (V_C) can be calculated as

$$\overline{V_C^2} = \int_0^\infty \frac{4kTR}{1 + (2\pi fRC)^2} = \frac{kT}{C}. \tag{3.20}$$

This same result can be obtained using the equipartition theorem. The RC circuit can be considered as a thermodynamically closed system with only one degree of freedom. The energy stored at the capacitor si given by $\frac{1}{2}CV^2$. Hence

$$\frac{1}{2}C\overline{V_C^2} = \frac{1}{2}kT, \tag{3.21}$$

which results in the same expression obtained in (3.20).

3.4.1.3 MOS Transistor Thermal Noise

Thermal noise in MOS transistors can be derived using the thermal noise model of a resistor described above. In the EKV model [18], the thermal noise is derived as follows. Each slice Δx of the MOS transistor channel is considered to have a local resistance ΔR. Its thermal noise is modeled by a current source ΔI_n with a PSD $4kT/\Delta R$. The contribution of this current noise source to the total drain current can be expressed as

$$\Delta S_{I_D} = G^2 \cdot \Delta R \cdot 4kT, \tag{3.22}$$

where G is the local transconductance associated to local interface charge density Q_i corresponding to the slice Δx. It is expressed as

$$G = \frac{\mu W |Q_i|}{L^2}, \tag{3.23}$$

with W and L being the gate width and length and μ, the carriers mobility.

The resistance of the channel slice Δx can be expressed as

$$\Delta R = \frac{\Delta x}{\mu W(-Q_i)}. \tag{3.24}$$

Hence, the total current noise PSD can be expressed as

$$S_{I_D} = 4kT\frac{\mu}{L^2}\int_0^L (-Q_i)dx = 4kT\frac{\mu Q_I}{L^2}, \tag{3.25}$$

where Q_I is the total inversion charge. Using the EKV formalism [18], the current PSD further simplifies to

$$S_{I_D} = 4kT\gamma G_m, \tag{3.26}$$

where G_m the gate transconductance of the transistor and γ is denoted the thermal excess noise factor and given by

$$\gamma = \begin{cases} \frac{1}{2}n \text{ in weak inversion and saturation,} \\ \frac{2}{3}n \text{ in strong inversion and saturation,} \end{cases} \tag{3.27}$$

where n is the EKV model parameter named the slope factor [18].

3.4.2 1/f and RTS Noise

The noise obeying the inverse frequency power law exists in practically all electronic and optoelectronic devices in addition to the thermal and quantum noise (shot noise). This noise has been well characterized over decades of cumulated experimental data. However, the exact mechanism of this low frequency noise is still a matter of debate and its exact origin is still not precisely identified. In electronic devices, the low frequency noise is denoted 1/*f* noise or flicker noise. The most common physical model of the 1/*f* noise relates its origin to defects resulting in alternate capture and emission of the charge carriers causing a fluctuation of the device conductance [19, 20] (Fig. 3.8).

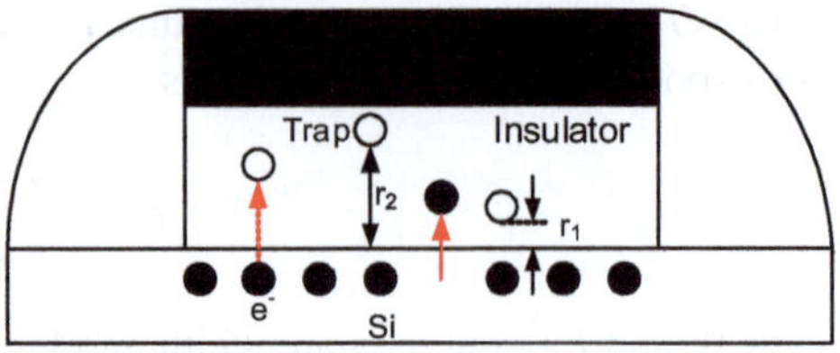

Fig. 3.8 Tunneling of charge carriers in the channel to the traps located in the insulator

For MOS transistors, the *1/f* noise physical models are historically based on two approches, the mobility fluctuation and the carrier number fluctuation [18]. The mobility fluctuation model has been first introduced by Hooge [21]. This model suggests that the *1/f* like fluctuation of the MOS device conductance is a result of the mobility fluctuation of the free carriers present in the device conducting channel. The carrier number fluctuation model has been first promoted by McWhorter [22]. It relates the *1/f* noise in MOS transistors to a process of generation-recombination between the carriers in the conduction band of the MOS channel and the traps located at the silicon oxide layer. A long debate lasted for years between the two schools of thought while other works tried to present theories reconciliating both models [23, 24].

Random telegraph signal (RTS) noise is the name commonly used to designate the fluctuations resulting from the process of capture and release of carriers by a single trap. This noise mechanism results in a discrete switching of the device conductance that manifests it self by switching events of the drain current between two or more states [19].

Advances in the process technologies allowed manufacturing tiny MOS transistors with deep-sub-micron gate widths and lengths featuring single traps. The study of tiny transistors featuring RTS noise showed that the drain current switching events correspond exactly to the carriers emission and release and bias voltage dependence of the this phenomenon allows the calculation of the distance separating the conducting channel from the traps in the insulator as well as their lateral position in the channel [25–28]. The depths of the traps have been estimated to be in the order of a few nm [25] which allows tunneling between the inversion layer and traps in the insulator.

The following is a brief review demonstrating that the RTS noise gives a plausible insight on the microscopic nature of the *1/f* noise.

The starting point is the modeling of the random switching process caused by a single trap. The MOS transistor drain current switches between two states denoted capture and release, respectively. These two states are characterized by two relaxation times denoted τ_c and τ_r. The probability density per unit time to switch from one state to the other is considered uniform and thus given by $\frac{1}{\tau_c}$ and $\frac{1}{\tau_r}$, respectively.

The autocorrelation of the switching process of the current is the key element to derive the noise PSD of the RTS noise caused by a single trap. Consider that the transitions between capture and release states cause the number of trapped carriers $N(t)$ to switch between 0 (release state) and 1 (capture state), Given that $N(t)$ can only take two possible values (0 and 1), the autocorrelation $R(t)$ of this random process can be expressed as:

$$R(t) = P_c \cdot P_{c,c}(t), \tag{3.28}$$

where P_c is the probability of being in the capture state at any given time, and $P_{c,c}(t)$ denotes the probability of being a the capture state at t given that the initial state is the same (An even number of transitions between times 0 and t).

We start by deriving $P_{c,c}(t)$. Based on the assumptions mentioned above, the probabilities to switch from capture and release states during dt are given by $\frac{dt}{\tau_c}$ and $\frac{dt}{\tau_r}$, respectively. Hence $P_{c,c}(t + dt)$ can be expressed as:

$$P_{c,c}(t + dt) = P_{c,c}(t)(1 - \frac{1}{\tau_c}) + (1 - P_{c,c}(t))\frac{1}{\tau_r}, \tag{3.29}$$

This differential equation leads the expression of $P_{c,c}(t + dt)$ as

$$P_{c,c}(t) = \frac{\tau_c}{\tau_r + \tau_c} + \frac{\tau_r}{\tau_r + \tau_c} exp\left(-\frac{\tau_c + \tau_r}{\tau_c \cdot \tau_r} t\right). \tag{3.30}$$

P_c can be derived in the same way by substituting P_c by $P_{c,c}(t)$ in (3.29). Note that P_c is time invariant, hence, in this case (3.29) simplifies to

$$P_c = \frac{\tau_c}{\tau_r + \tau_c}. \tag{3.31}$$

The autocorrelation can be derived fron (3.28), (3.30) and (3.31) and used with the wiener khintchine theorem to derive the RTS noise PSD as

$$S_{RTS}(f) = 4\int_0^{\infty} R(t)cos(2\pi ft)dt = P_c^2 \frac{4\tau_c}{1 + (2\pi f \tau_c)^2} \tag{3.32}$$

The probability that single trap is in the capture or release states can be calculated using the grand partition function leading to the expression of the ratio between capture and release rates as

$$P_c = 1 - \left(1 + g \cdot exp(\frac{E - E_F}{kT})\right)^{-1}, \tag{3.33}$$

where g is the degeneracy factor of the trap, E is the energy that marks the Fermi level E_F at which the trap makes the transition to the capture state.

τ_c depends, at a given temperature, on the depth of the trap in the oxide and the energy of the trap. The PSD of the noise caused by all the traps in the silicon oxide is obtained by integrating (3.32) over all the energies E and τ_c in the gate oxide volume. The carrier capture is supposed to occur by a charge tunneling from the inversion layer of the MOS transistor to the trap through an insulator (silicon oxide). Hence, if the trap is located at a distance r from the interface, the capture time constant τ_c can be expressed as [29]:

$$\tau_c = \tau_0 \cdot e^{\frac{r}{\lambda}}, \tag{3.34}$$

where τ_0 is considered to be a constant and λ is the tunneling attenuation distance ($\simeq$0.1 nm) [30].

It is not easy to determine the spatial distribution of traps with a given energy in the oxide [31]. In order to give a simple example, we consider that the traps are uniformly distributed between depths r_1 and r_2 in the oxide corresponding to the capture time constants τ_{c1} and τ_{c2}. We also consider a uniform distribution over a range of possible trap energies. In this simplified model, the total noise resulting from the traps located between the depths r_1 and r_2 in the oxide has a PSD density per unit area of the gate oxide that can be derived by integrating (3.32) over the trap energies E and depths r which results in:

$$S_{\Sigma RTS}(f) = k \cdot T \cdot \lambda \cdot N_t \cdot \frac{2}{\pi} \cdot \frac{arctan(2\pi f \tau_2) - arctan(2\pi f \tau_1)}{f} \simeq \frac{k \cdot T \cdot \lambda \cdot N_t}{f}. \tag{3.35}$$

where N_t is a constant in $(eV^{-1} \cdot m^{-3})$ including all the constants of integration. N_t is referred to as the oxide trap density, it is obtained by measurements and takes values in the range of 10^{16} to 10^{17} $eV^{-1} \cdot cm^{-3}$ at room temperature. The effect of the *arctan* function is very weak for frequencies verifying $\frac{1}{\tau_{c1}} \ll f \ll \frac{1}{\tau_{c2}}$. Which results in a noise PSD proportional to $1/f$. But when the frequency tends to zero, the *arctan* term sets the PSD to a constant value instead of diverging to infinity if only a $1/f$ behavior is assumed, which is physically more plausible.

The trapping mechanism discussed above results in a fluctuation of the number of charges in the channel and also a variation of the surface potential in the MOS transistor channel. For a MOS transistor with a gate width W and length L, the PSD of the noise resulting from the fluctuation of the number of carriers in the channel can be calculated from (3.35) as

$$S_N(f) \simeq \frac{k \cdot T \cdot \lambda \cdot N_t \cdot W \cdot L}{f}. \tag{3.36}$$

It is generally agreed that this variation of the number of trapped charges induces correlated fluctuation of the channel carrier number and mobility [20]. The derivation of the drain current noise associated with these variations is performed by relating the drain current variation to the channel carriers mobility and number fluctuation. Both the Berkeley Short-channel IGFET Model (BSIM), and EKV [18, 23] formalisms lead to equivalent gate referred $1/f$ noise expressions. In the EKV model, it is expressed as

$$S_{V_g}(f) = \frac{q^2 \cdot k \cdot T \cdot \lambda \cdot N_t \cdot K_G}{C_{ox}^2 \cdot W \cdot L \cdot f}, \tag{3.37}$$

where K_G is a bias dependent parameter close to unity when the transistor is operating in the weak and moderate inversion regime. Otherwise, it increases with the inversion coefficient. In the BSIM model N_t is denoted A or $NoiA$.

3.4.3 Leakage Current Shot Noise

3.4.3.1 Current Shot Noise

The shot noise originates from the quantized nature of charge constituting the current. For electronic devices, it has been first revealed in vacuum tubes by Walter Schottky in 1918. The shot noise appears as a fluctuation of a steady current flow crossing a barrier between two regions (materials). This steady current is the result of independent individual and discrete charge carriers motion. The shot noise does not appear in resistive devices since the charge carriers are not independent form each other. As discussed previously for the photon shot noise. This noise is the result of the fact that the number of particles crossing the barrier between two regions, per unit time, behaves as a Poisson process. In other words, the variance of the fluctuation of the number of particles crossing the barrier per unit time is simply given by their average number.

The current flowing through the device depicted in Fig. 3.9 is a series of discrete events. Charge carriers cross the barrier between the regions of the device at random instants t_i. Figure 3.10a depicts the evolution, in time, of the number of cumulated charge carriers crossing the barrier $n_c(t)$. $n_c(t)$ shows a behavior similar to the "drunken man's walk". In a time interval $[t; t + h[$, the number of charges crossing the barrier is expressed as

$$n(t, h) = n_c(t + h) - n_c(t), \tag{3.38}$$

Since the number of charges crossing the barrier during a given time interval is a Poisson process as discussed above, the variable $n(t, h)$ verifies the following properties:

First, for a given (t, h), the average and variance of the variable $n(t, h)$ are equal and depend only on the duration h. These statistical values can be related to the mean (dc) current crossing the device I_0 as

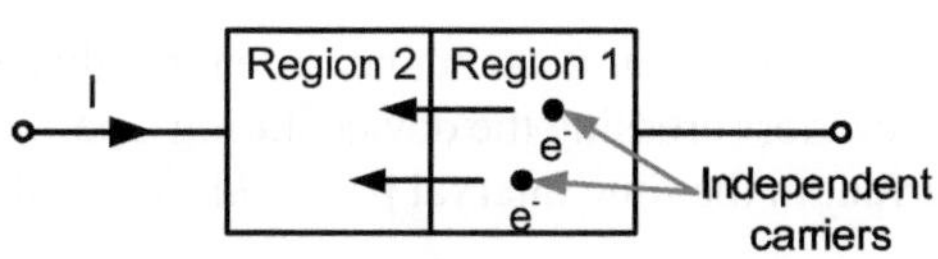

Fig. 3.9 Typical example of a device featuring current shot noise. The individual charge carriers constituting the current I cross the barrier between the two regions of the device at random moments

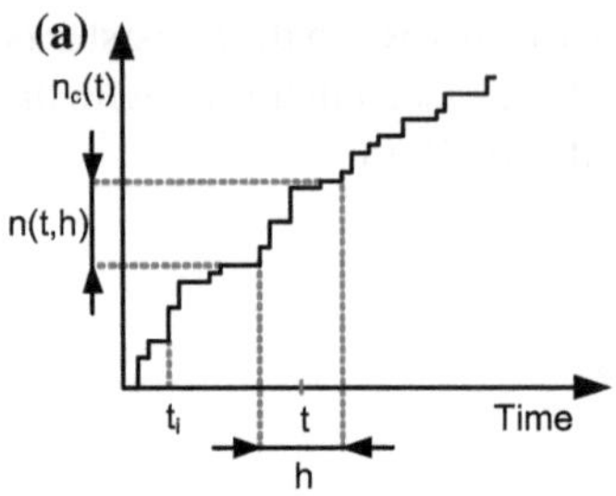

Fig. 3.10 **a** The evolution, in time, of the number of cumulated charge carriers crossing the barrier $n_c(t)$. $n_c(t)$ shows a behavior similar to the "drunken man's walk". **b** The current measured at t as the number of charges crossing the device in the time interval $[t, t + h[$, $n(t, h)$, divided by h

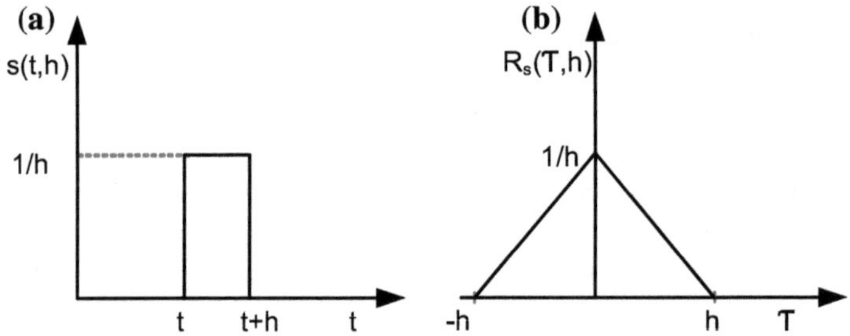

Fig. 3.11 The square function $s(t, h)$ used to define the measured current at t by counting the charges crossing the device in the interval $[t, t + h[$ (**a**) and the autocorrelation function corresponding to $s(t, h)$) (**b**)

$$Var[n(t, h)] = E[n(t, h)] = h \cdot \frac{I_0}{q}. \tag{3.39}$$

The second property is related to the independence of the charge carriers from each other. This implies that the number of charge carriers crossing the barrier in a time interval $[t; t + h]$ is uncorrelated with the number crossing it during $[t + \tau; t + \tau + h]$ if τ is higher than h.

The current crossing the device at t is ideally given by

$$I(t) = q \cdot \lim_{h \to 0} \frac{n(t, h)}{h}. \tag{3.40}$$

Practically, the current at an instant t can be measured in a discrete way by counting the number of charge carriers crossing the device during a period h as $\frac{n(t,h)}{h}$ as shown in Fig. 3.10b. The current in the time interval $[t; t + h[$, denoted $I(t, h)$, is then given by

$$I(t, h) = q \cdot n(t, h) \cdot s(t, h), \tag{3.41}$$

where $s(t, h)$, shown in Fig. 3.11a, is the square function taking $1/h$ between t and $t + h$ and zero elsewhere. We are now interested at deriving the autocorrelation of the current $I(t, h)$ measured in the interval $[t; t + h[$. The autocorrelation can be calculated as

$$R_I(\tau, h) = q^2 \cdot E[\int_{-\infty}^{+\infty} n(t, h)s(t, h)s(t + \tau, h)dt] = q^2 \cdot \int_{-\infty}^{+\infty} E[n(t, h)]s(t, h)s(t + \tau, h)dt. \quad (3.42)$$

Using (3.39), this equation simplifies to

$$R_I(\tau, h) = q \cdot I_0 \cdot h \cdot R_s(\tau, h), \quad (3.43)$$

where $R_s(\tau, h)$ is the autocorrelation function of $s(t, h)$ depicted in Fig. 3.11b. Finally the noise PSD of the current $I(t, h)$ can be calculated using the Wiener-Khintchine theorem as the Fourier transform of $R_I(\tau, h)$. Knowing that

$$\mathscr{F}\{R_s(\tau, h)\}(f) = \frac{1}{h} \cdot (sinc(\pi f h))^2, \quad (3.44)$$

where $sinc(x) = \frac{sin(x)}{x}$. The bilateral PSD of $I(t, h)$ fluctuation is expressed using (3.43) and (3.44) and multiplying by 2 as

$$S_I(f, h) = 2 \cdot q \cdot I_0 \cdot (sinc(\pi f h))^2. \quad (3.45)$$

As the current $I(t)$ is defined as the limit when h tends to 0, one can conclude that PSD of the ideal current $I(t)$ tends to a white PSD expressed as

$$S_I(f) \simeq 2 \cdot q \cdot I_0. \quad (3.46)$$

3.4.3.2 Leakage Current Shot Noise

The leakage is the charge transport observed in depleted regions or through insulators in electronic devices. In the context of CIS, the most sensitive part to such a phenomena is the sense node. At the level of the pixel sense node, one can distinguish two main leakage current mechanisms depicted in Fig. 3.12. The first originates from the generation recombination mechanisms in the depleted region of the pn junctions connected to the sense node and particularly the floating diffusion. This leakage current mechanism is identical to the one behind the dark current generation in the depleted region of PPD. The second mechanism behind the sense node leakage current is the charge tunneling through the gate dielectric of the readout transistor. Indeed, the thickness of the gate oxide in modern CMOS processes gives rise to this phenomenon that finds its physical model in quantum mechanics.

The junction leakage current flows between the sense node and the bulk connected to the ground. The tunneling current separates in three components flowing between

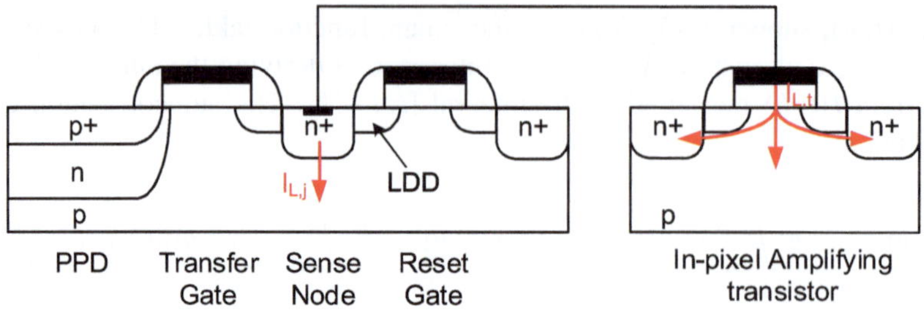

Fig. 3.12 The leakage current sources taking place at the level of the sense node of a CIS pixel

the gate and each one of the readout transistor terminals, namely, the source, the drain and the bulk.

Each of these leakage current mechanisms corresponds to individual charge carriers crossing a barrier between two materials. Hence they result in a shot noise as described above. Consequently, the shot noise PSD and variance are directly linked to the leakage current mean value. Quantifying the average current allows calculating the variance of the leakage current shot noise.

The simulated average leakage current at the level of the sense node for technology nodes above 130 nm is below 0.001 e^- for a time lapse between the reset and the transfer samples of 10 μs. Hence, for these technology nodes, the leakage current shot noise is complectly negligible. The leakage current associated to the tunneling through the in-pixel amplifying transistor depends exponentially on the gate oxide thickness. Hence for more advanced technologies, the leakage current increases by several orders of magnitude and could become a limiting factor for noise. This point will be discussed in Chap. 9.

3.5 Fixed Pattern Noise

The noise sources presented above are random fluctuations in time. The fixed pattern noise (FPN) is rather a spatial variation over the pixels array. The spatial variation is random from chip to chip but correlated in time for each chip. Hence It is a noise that can be compensated by means of image post processing. The main origins of FPN are pixel-to-pixel photo response non uniformity (PRNU), column level gain variation and offset spatial variation. The gain variation dominates in the presence of a high signal level and the offset variations dominate the FPN in low light conditions. Hence, the FPN can be categorized in spatial gain variation and offset spatial variation.

Consider an image sensor pixel array exposed to a uniform light intensity. The signal at the output of each pixel can be expressed as

$$V_{\text{out}}(t) = A_{\text{CG}} \cdot N(t) + V_{TRN}(t) + V_{off}, \tag{3.47}$$

where A_{CG} is the overall conversion gain of the readout chain, $N(t)$ the number of photons integrated by the pixel PPD and transferred to its sense node. $V_{TRN}(t)$ is the temporal read noise, V_{off} is the residual offset voltage (constant in time), remaining after the CDS, that can originate for instance from the charge injection caused by the transfer gate at the level of the sense node. Let's consider that each pixel is read several times for the same amount of light intensity corresponding to an average of N photoelectrons. In this case the temporal average of the signal at the output of each pixel cancels the temporal read noise and the temporal average output signal of each pixel can be expressed as

$$E_t[V_{\text{out}}] = A_{\text{CG}} \cdot N + V_{off}, \tag{3.48}$$

where $E_t[]$ denotes the average over time for a single pixel.

The fixed pattern noise originates from the spatial variation of A_{CG} under an amount of light high enough and it is dominated by the spatial variation of V_{off} under low light conditions. The spatial variance over the pixels array of the temporal average of the signal is expressed as:

$$Var_s[E_t[V_{\text{out}}]] = \frac{Var_s[A_{\text{CG}}]}{A_{\text{CG}}^2} \cdot E_{st}[V_{\text{out}}]^2 + Var[V_{off}], \tag{3.49}$$

where $Var_s[]$ denotes the spatial variance (variations over the pixels) and $E_{st}[]$ denote the average over time and space.

This equation can be used to characterize the FPN in CIS. the characterization requires several image cubicals obtained at different exposure times or lighting levels. Each cubical consists in consecutive images while all the pixels are uniformly exposed to the source of light. The cubical allows computing the average values over time for each pixel and then the spatial variance over the pixels. A single variance and average values are then obtained for each cubical. Representing the variance values obtained for different exposures as a function of the corresponding squared average values yields in a linear curve determining the gain variance by its slope and the offset variance by its origin. Chapter 7 shows a practical example on a VGA imager.

3.5.1 Spatial Gain Variation

The spatial gain standard deviation $\sigma_{A_{\text{CG}}}$ is expressed as a percentage of the average conversion gain as

$$\sigma_{A_{\text{CG}}} = \sqrt{\frac{Var[A_{\text{CG}}]}{A_{\text{CG}}^2}}. \tag{3.50}$$

This spatial variation is due to the pixel-to-pixel non uniformity and column level gain non uniformity.

The vertical gain mismatch originates from the gain variation of the column level amplifiers and results in vertical lines over the image. It is very easy to recognize for the human eye which makes it critical for the image quality. The column level gain is generally implemented with switched capacitor amplifiers. Hence this gain spatial variation is mainly due to capacitors mismatch. In order to implement high gain levels, small feedback capacitors are needed which make their mismatch the main source of vertical spatial noise. Typical values of this mismatch standard deviation are of the order of 0.1% [32].

The pixel-to-pixel non uniformity is not easy to recognize with the human eye for a single image since it cannot be easily distinguished from the photon shot noise and the temporal read noise. The pixel-to-pixel non-uniformity can have multiple origins. the first comes from the PPD that may suffer from a spatial variation of the quantum efficiency, pin voltage or the full well capacity. These effects dominate at high levels of illumination. The second source is the conversion gain mismatch. The latter can be the result of the sense node capacitance variation which can be due to the spatial variation of sense node junction, the reset at transfer gates overlap capacitances with the sense node and the source follower transitory intrinsic and extrinsic [18] parameters. Typical values of this standard deviation are of the order of 0.5% [32].

3.5.2 *Offset Variations*

The main sources of offset are the transistors located in the pixel and the column level amplifier. The column level amplifier auto-zeroing and the double sampling readout scheme reduce efficiently the offset of the in-pixel source follower stage as well as the column level amplifier. Offset can also be related to the nonidealities of the switches connected to the sense node. Indeed the transfer gate and reset switches introduce a clock feedthrough resulting in a charge injection into the sense node. The injected charge ΔQ can be generally expressed as

$$\Delta Q = \frac{C_{ov} \cdot C_{SN}}{C_{ov} + C_{SN}} \Delta V_G \simeq C_{ov} \Delta V_G, \tag{3.51}$$

where C_{SN} denotes the total sense node capacitance. C_{ov} is the overlap or coupling capacitance between the transfer or reset gates and the sense node. ΔV_G is the amplitude of pulse applied at the level of the reset of transfer gates. For the transfer gate, the charge injected at sense node when the transfer gate voltage switches between low and high level is compensated by the one injected when the latter goes back from high to low level. The charge injected after the reset is canceled with the sampled reset noise thanks to the double sampling scheme. In the case of a high in pixel gain (common source based configuration) [33], the charge injected in the sense node after the reset can introduce an offset even with CDS. Indeed, the spatial variation of the reset switch may cause a spatial variation of the sense node reset level witch results in pixel gain variation.

3.6 Discussion

The different noise sources presented in this chapter affect the CIS SNR differently depending on the number of input photons and consequently on the lighting conditions of the image. In order to illustrate this effect, we calculate a simplified formula of the CIS readout chain SNR. The latter can be expressed as

$$SNR(N) = 10Log\left[\frac{N^2}{\sigma_{Sh}^2 + \sigma_D^2 + \sigma_T^2 + \sigma_F^2 + \sigma_R^2}\right]. \tag{3.52}$$

N denotes the average number of photoelectrons generated in the PPD. The denominator of this expression includes all the noise sources discussed in the previous sections where each is presented by an input-referred standard deviation expressed in e^-_{rms}. σ_{Sh} presents the photon shot noise standard deviation which is expressed as

$$\sigma_{Sh}^2 = N. \tag{3.53}$$

σ_D presents the dark current shot noise

$$\sigma_D^2 = N_D. \tag{3.54}$$

where N_D is the average number of electrons generated by the dark current in the PPD for a given integration time.

σ_T presents the standard deviation of the transfer noise. The transfer noise is expressed as a shot noise related to the charge transfer inefficiency (CTI) also referred to as the lag. Hence

$$\sigma_T^2 = CTI \cdot N. \tag{3.55}$$

σ_F represents the standard deviation of the input-referred FPN. We consider the offset spatial noise negligible (dark fixed pattern noise). σ_F can be expressed as

$$\sigma_F^2 = \sigma_{A_{CG}}^2 \cdot N^2. \tag{3.56}$$

Finally, σ_R represents the total input-referred read noise, expressed in e^-_{rms}, including the thermal, 1/*f* and shot noise causing random fluctuations in the electronic readout circuitry.

Figure 3.13 shows in dB the level of the signal as well as the different noises presented above as a function of the input number of photoelectrons for typical values of the dark current of 0.1 e^-_{rms}, a CTI of 0.1%, a FPN corresponding to a $\sigma_{A_{CG}}$ of 1%. Figure 3.13 shows that the signal level depending noises, namely the photon shot noise and the FPN dominate the noise at high signal level. Indeed the photon shot noise dominates in the mid range signal level and the FPN in higher signal levels if the sensor is not already saturated. At low signal level, the importance of a low readout noise is shown by the curves corresponding to the two read noise levels of 10 and 1 e^-_{rms}. In order to compare with respect to the signal, Fig. 3.14 shows the SNR as

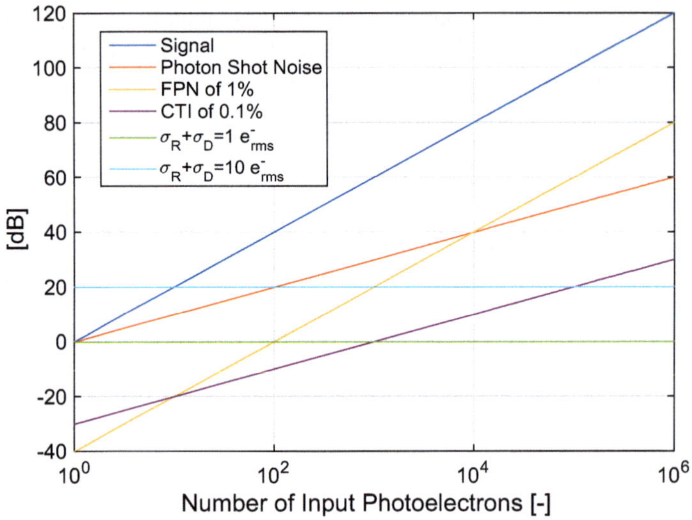

Fig. 3.13 Signal and different noises level as a function of the number of photoelectrons

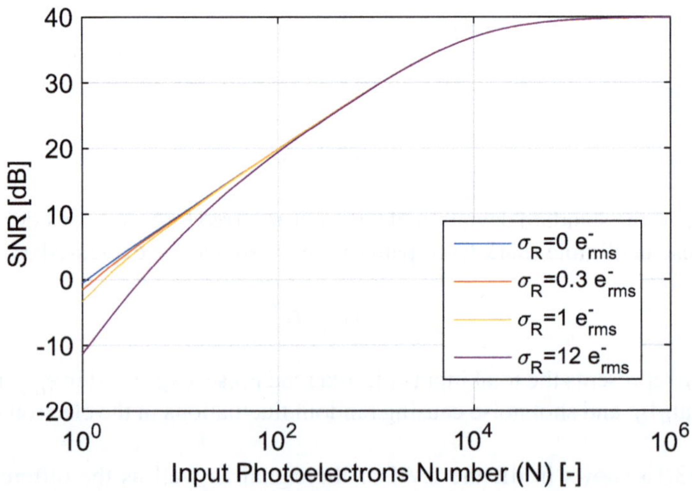

Fig. 3.14 Evolution of the CIS readout chain SNR as a function of the number of photoelectrons

a function of the number of input photoelectrons for a dark current of 0.1 e^-_{rms} for the integration time, a CTI 0.1%, a FPN corresponding to a $\sigma_{A_{CG}}$ of 1% and a read noise from the ideal $\sigma_R^2 = 0\,e^-_{rms}$ to $\sigma_R^2 = 12\,e^-_{rms}$. From Fig. 3.14, one can distinguish, as mentioned above, three areas of noise dominance. At low light level, the read noise appears as the dominant noise source and the SNR is sensitive to the read noise only in this region. For a number of input photoelectrons about ten times higher than the read noise level, the photon shot noise dominates. Thus, CIS featuring extremely low read noise levels are called shot noise limited image sensors. Such condition becomes valid for a read noise in the subelectron level as shown by Fig. 3.14. When the imager is exposed to a high light intensity, the FPN appears as the dominant noise source limiting the SNR if the sensor is not already saturated.

The above discussion shows that the electronic read noise is the bottleneck for ultimate low-light performance in CIS. The following Chapter will focus on the detailed analytical evaluation of the term σ_R.

References

1. A. Papoulis, *Probability, Random Variables, and Stochastic Processes* (McGraw-Hill, New York, 1965), pp. 345–396
2. J.R. Janesick, *Photon Transfer* (SPIE, USA, 2007), pp. 49–79
3. J.R. Janesick, *Scientific Charge-Coupled Devices* (SPIE, USA, 2001), pp. 605–649
4. C.T. Sah, R.N. Noyce, W. Shockley, Carrier generation and recombination in p-n junctions and p-n junction characteristics. Proc. IRE **45**(9), 1228–1243 (1957)
5. R.N. Hall, Electron-hole recombination in germanium. Phys. Rev. **87**, 387 (1952). https://doi.org/10.1103/PhysRev.87.387
6. W. Shockley, W.T. Read, Statistics of the recombinations of holes and electrons. Phys. Rev. **87**, 835–842 (1952). https://doi.org/10.1103/PhysRev.87.835
7. S. Sze, K. Ng, *Physics of Semiconductor Devices* (Wiley, Hoboken, 2006), pp. 7–79, https://books.google.ch/books?id=o4unkmHBHb8C
8. V. Goiffon, C. Virmontois, P. Magnan, Investigation of dark current random telegraph signal in pinned photodiode cmos image sensors, in *2011 IEEE International on Electron Devices Meeting (IEDM)* (2011), pp. 8.4.1–8.4.4
9. X. Wang, P.R. Rao, A.J.P. Theuwissen, Fixed-pattern noise induced by transmission gate in pinned 4t cmos image sensor pixels, in *Proceeding of the 36th European Solid-State Device Research Conference, 2006. ESSDERC 2006* (2006), pp. 331–334
10. B. Mheen, Y.J. Song, A.J.P. Theuwissen, Negative offset operation of four-transistor cmos image pixels for increased well capacity and suppressed dark current. IEEE Electron Device Lett. **29**(4), 347–349 (2008)
11. H. Han, H. Park, P. Altice, W. Choi, Y. Lim, S. Lee, S. Kang, J. kim, K. Yoon, J. Hynecek, Evaluation of a small negative transfer gate bias on the performance of 4t cmos image sensor pixels, in *International Image Sensors Workshop* (2007), pp. 238–240
12. E.R. Fossum, Charge transfer noise and lag in cmos active pixel sensors, in *Proceedings of 2003 IEEE Workshop on Charge-Coupled Devices and Advanced Image Sensors* (2003), pp. 1–6
13. K. Thornber, Noise suppression in charge transfer devices. Proc. IEEE **60**(9), 1113–1114 (1972)
14. J.R. Janesick, *Scientific Charge-Coupled Devices* (SPIE, USA, 2001), p. 656
15. L.E. Bonjour, N. Blanc, M. Kayal, Experimental analysis of lag sources in pinned photodiodes. IEEE Electron Device Lett. **33**(12), 1735–1737 (2012)

16. A. Boukhayma, A. Peizerat, C. Enz, A 0.4 e-rms temporal readout noise 7.5 m pitch and a 66% fill factor pixel for low light cmos image sensors, in *International Image Sensors Workshop* (2015)
17. A. Boukhayma, A. Peizerat, C. Enz, Temporal readout noise analysis and reduction techniques for low-light cmos image sensors. IEEE Trans. Electron Devices **63**(1), 72–78 (2016)
18. C. Enz, E. Vittoz, *Charge based MOS Transistor Modeling, The EkV Model for Low-power And RF IC Design* (Wiley, Hoboken, 2006)
19. C. Surya, T.Y. Hsiang, Surface mobility fluctuations in metal-oxide-semiconductor field-effect transistors, Phys. Rev. B **35**, 6343–6347 (1987). https://doi.org/10.1103/PhysRevB.35.6343
20. K.K. Hung, P.K. Ko, C. Hu, Y.C. Cheng, Random telegraph noise of deep-submicrometer mosfets. IEEE Electron Device Lett. **11**(2), 90–92 (1990)
21. F.N. Hooge, 1/f noise sources. IEEE Trans. Electron Devices **41**(11), 1926–1935 (1994)
22. A. McWhorter, 1/f noise and Germanium surface properties, in *Semiconductor Surface Physics*. Series in Physics (University of Pennsylvania Press, 1957), pp. 207–228
23. K. Hung, P. Ko, C. Hu, Y. Cheng, A physics-based mosfet noise model for circuit simulators. IEEE Trans. Electron Devices **37**(5), 1323–1333 (1990)
24. G. Ghibaudo, O. Roux-dit Buisson, J. Brini, Impact of scaling down on low frequency noise in silicon mos transistors, physica status solidi (a) **132**(2), 501–507 (1992). https://doi.org/10.1002/pssa.2211320226
25. N.V. Amarasinghe, Z. elik Butler, P. Vasina, Characterization of oxide traps in 0.15 μm^2 mosfets using random telegraph signals. Microelectron. Reliab. **40**(11), 1875–1881 (1999)
26. Z. Celik-Butler, P. Vasina, N.V. Amarasinghe, A method for locating the position of oxide traps responsible for random telegraph signals in submicron mosfets. IEEE Trans. Electron Devices **47**(3), 646–648 (2000)
27. T. Nagumo, K. Takeuchi, T. Hase, Y. Hayashi, Statistical characterization of trap position, energy, amplitude and time constants by rtn measurement of multiple individual traps, in *2010 IEEE International Electron Devices Meeting (IEDM)* (2010), pp. 28.3.1–28.3.4
28. V. Sedlakova, J. Sikula, M. Chvatal, J. Pavelka, M. Tacano, M. Toita, Noise in submicron metaloxidesemiconductor field effect transistors: lateral electron density distribution and active trap position. Jpn. J. Appl. Phys. **51**(2R), 024105 (2012), http://stacks.iop.org/1347-4065/51/i=2R/a=024105
29. I. Lundstrm, C. Svensson, Tunneling to traps in insulators. J. Appl. Phys. **43**(12), 5045–5047 (1972). https://doi.org/10.1063/1.1661067
30. G. Ghibaudo, Low-frequency noise and fluctuations in advanced cmos devices. Proc. SPIE **5113**, 16–28 (2003). https://doi.org/10.1117/12.484913
31. Y. Nemirovsky, D. Corcos, I. Brouk, A. Nemirovsky, S. Chaudhry, 1/f noise in advanced cmos transistors. IEEE Instrum. Meas. Mag. **14**(1), 14–22 (2011)
32. D. Sackett, Cmos pixel device physics, in *IEEE Solid State Circuits Conference on Image Sensor Design Forum: Charachterization of Solid-State Image Sensors, San Francisco, CA, USA* (2005)
33. C. Lotto, P. Seitz, T. Baechler, A sub-electron readout noise cmos image sensor with pixel-level open-loop voltage amplification, in *2011 IEEE International Solid-State Circuits Conference* (2011), pp. 402–404

Chapter 4
Detailed Noise Analysis in Low-Noise CMOS Image Sensors

Based on the fundamental background of the different noise sources presented in the previous Chapter, a noise analysis for CIS readout chains is presented. It starts by a preamble recapitulating the impact of the different noise sources on the signal path of a CIS and pointing-out, based on quantitative values from state-of-the-art low noise CIS, the electronic readout noise as the main contributor. A detailed analysis of the $1/f$ noise, thermal noise, and leakage current shot noise is then presented for readout chains based on the two possible pixel configurations, namely, the common source and source follower topologies. These two configurations are then compared.

4.1 Preamble

Figure 4.7 shows the schematic of a conventional low-noise CIS readout chain. The corresponding timing diagram is shown in Fig. 4.1. It also shows the potential profile across the PPD, the transfer gate (TX) and the sense node (SN) during the three phases of operation: the integration, the reset and the transfer phases. During the integration time, the PPD accumulates the electrons generated by the incident photons. During the readout, the pixel is connected to the column through the row selection switch (RS) then the reset switch (RST) is closed in order to set the SN voltage higher than the pinning voltage of the PPD. The voltage level at the SN after the reset is read with the in-pixel source follower (SF) and sampled onto the end of the readout chain. The potential barrier between the PPD and the SN is controlled by the transfer gate (TX). When the barrier is lowered, the charges accumulated in the PPD are transferred to the SN. The SN voltage level after the transfer is sampled at the output of the readout chain. The reset and transfer samples are then differentiated. This operation is called correlated double sampling (CDS) [1].

Figure 4.1 depicts also the different noise sources affecting the signal in the CIS apart from the photon shot noise. During the integration, the charge originating from

A. Boukhayma, *Ultra Low Noise CMOS Image Sensors*,
Springer Theses, https://doi.org/10.1007/978-3-319-68774-2_4

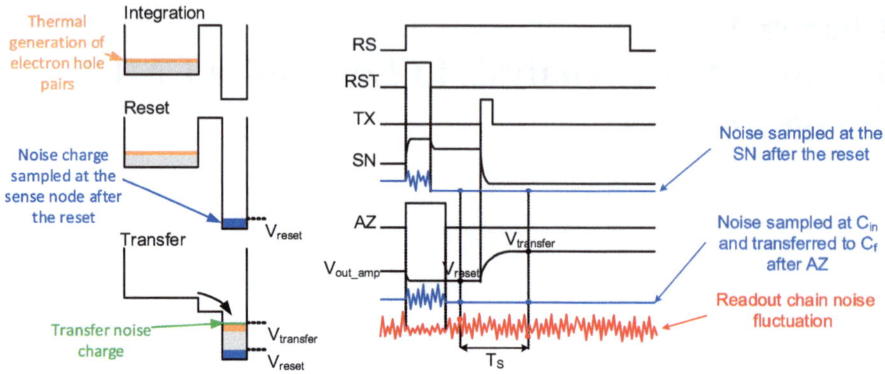

Fig. 4.1 Timing diagram of the conventional CIS readout chain with noise mechanisms affecting the signal at the PPD and the readout chain levels

thermal generation of electron-hole pairs in the depleted region of the PPD (the dark current) can corrupt the signal. In state-of-the-art CIS, the dark current in PPDs has been reduced to a few e^{-}/s. Hence, for exposure times below hundreds of ms the dark current can be neglected.

The reset of the SN leaves a kTC noise charge held at the SN capacitance. This noise is as high as several electrons in the case of a SN capacitance of a few fF. But for 4T pixels, it is cancelled thanks to the CDS readout scheme as depicted in the timing diagram of Fig. 4.1.

The charge transfer from the PPD to the SN can be affected by the noise related to the charge deficit due to incomplete transfer and lag [2]. Unlike the sampled reset kTC noise, this noise is not cancelled by the CDS. The charge transfer noise has been extensively studied for CCDs [3, 4] because an efficient charge transfer is crucial in such devices. But this issue has not been discussed as much for the case of 4T pixels [2]. This is likely due to the fact that only one transfer is needed, and the noise is believed to be dominated by the photon shot noise, at high signal level, or by the read noise at low light. In state-of-the-art CIS with 4T pixels, values of lag as low as 0.1% have been reported. Thus the lag can be neglected compared to the read noise in the low light context. The transient noise related to the lag is believed to behave as a shot noise [2] similarly to buried channel CCDs [4]. But with a lag below 1%, this noise can be neglected in low light conditions.

Finally, the readout of the SN reset and transfer voltages is affected by random fluctuations due to the readout chain noise. Starting with the in-pixel SF and noise coupling of the (TX) and (RST) lines with the SN, the power supply noise, and ending with the column level circuitry and analog-to-digital converters (ADCs). The column level amplification is introduced in order to minimize the contribution of the next circuit blocks to the input-referred total noise, e.g. buffers, sample-and-holds, and ADC. The column level amplifier also limits the bandwidth in order to minimize the thermal noise [5]. A switched capacitor amplifier is usually used. An auto-zero

(AZ) is performed in order to reset its feedback capacitor and reduce its offset and $1/f$ noise [1]. When the AZ switch is opened, the noise is sampled at the integration capacitor and transferred to the output. This sampled noise is also canceled thanks to the CDS since it is common to the reset and transfer samples as depicted in Fig. 4.1. Low noise CIS readout chains may also include correlated multiple sampling (CMS) that can be implemented with analog circuitry [6, 7] or performed in the digital domain after the ADC [8]. CMS consists in averaging M samples after the reset and M other samples after the transfer with a sampling period T_S, then calculating the difference between the two averages.

With careful design, the readout noise originating from the pixel and column-level amplifier are the dominant noise sources in CIS. A detailed noise analysis of the readout noise is therefore necessary in order to determine the key design and process parameters that can be used for further noise reduction.

It has been shown, in the previous section, that among the different noise sources in CIS readout chains, the read noise originating from the MOS transistors of the readout chain remains the bottleneck for ultimate sensitivity. In this Chapter, a detailed noise analysis is presented. The analysis comprises modeling as signal processing blocks of the different stages of the CIS readout chain, namely, the pixel, the column level amplification and CMS. The analytical calculation leads to simple equations of the input referred $1/f$ noise, thermal noise, and leakage current shot noise. The Chapter starts by detailing the noise calculation methodology, then the impact of CMS on noise is studied in order to derive the transfer function of CMS. Then, two pixel configurations are studied and compared (from the noise perspective), namely the source follower based pixel and the common source (CTIA) based pixel.

4.2 Noise Calculation Method

In the following analysis, it is assumed that the noise sources of the different devices are statistically independent and hence uncorrelated. The transfer function from each noise source to the output is first calculated. The total output referred noise variance is then given by the sum of all variances corresponding to each noise source. The $1/f$ noise, thermal noise, and leakage current shot noise are analyzed separately in order to clarify their relative impact on the output-referred noise.

In low noise CIS readout chains, column level amplification is generally implemented between the pixel stage and the next stages like CDS/CMS and ADC. With enough column-level gain and careful design the kTC noise of the sampling stages and the quantization noise of the ADC can be made negligible compared to the noise originating from the pixel and amplification stages. The reset kTC noise sampled at the SN and the noise sampled at the input capacitor of the column-level amplifier after auto-zeroing and transferred to the output can be canceled efficiently thanks to the double sampling scheme of CIS readout chains. Hence, only two dominant noise sources are considered, namely, one from the in-pixel amplifying transistor, and one from the column amplifier. The noise originating from the bias circuits and from the

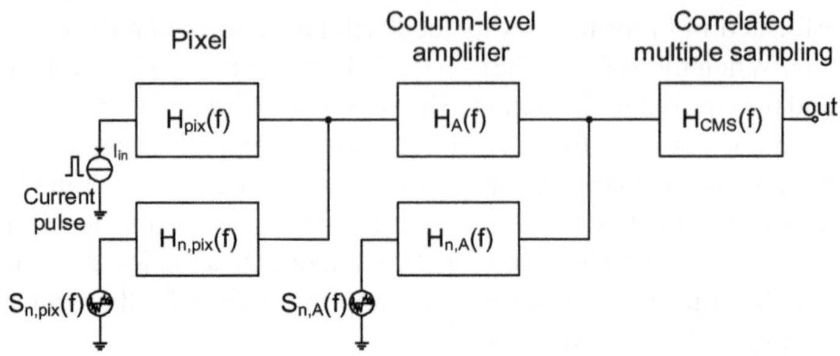

Fig. 4.2 Signal processing block diagram used for the noise analysis showing the signal and noise paths

power supply can be included into these two noise sources. For the $1/f$ noise, the column-level amplifier contribution is not included since it can be made negligible by designing the column-level saturated transistors to have gate areas much larger than the in-pixel amplifying transistor.

Figure 4.2 shows the signal processing block diagram used for the noise analysis. The input signal of the CIS readout chain is the charge transferred to the sense node. The PPD is hence modeled as a current source pulse injecting the charge accumulated during the integration phase in the sense node capacitance when the transfer gate is opened. The signal transfer function is first calculated in order to determine the overall conversion gain of the readout chain. The pixel signal transfer function relating the column level voltage to that charge pulse is denoted $H_{\text{pix}}(f)$, its magnitude corresponds to the pixel conversion gain denoted A_{CG}. The transfer functions of the column-level amplifier and the correlated multiple sampling are denoted respectively $H_A(f)$ and $H_{\text{CMS}}(f)$ respectively. For the readout circuit, the noise mechanism is better described as a current or voltage fluctuation. Thus, the readout noise is calculated at the output of the signal path as a voltage using the noise transfer functions, then referred to the input as a charge after division by the gain of the signal path, namely the pixel conversion gain A_{CG} and column level amplifier gain A_{col}. Thus, the variance of the input-referred noise originating from the pixel can be expressed as

$$\overline{Q_{\text{pix}}^2} = \frac{1}{A_{\text{CG}}^2 \cdot A_{\text{col}}^2} \cdot \int_0^{\infty} S_{\text{n,pix}}(f) \cdot |H_{\text{n,pix}}(f)|^2 \cdot |H_A(f)|^2 \cdot |H_{\text{CMS}}(f)|^2 df, \quad (4.1)$$

where $S_{\text{n,pix}}(f)$ refers to the PSD of the current noise source modeling the noise sources at the pixel-level readout circuitry, and $H_{\text{n,pix}}(f)$ represents the transfer function from that source to the column voltage.

Similarly, the variance of the input-referred noise originating from the column-level amplifier circuitry can be expressed as

$$\overline{Q_A^2} = \frac{1}{A_{\mathrm{CG}}^2 \cdot A_{\mathrm{col}}^2} \cdot \int_0^{\infty} S_{\mathrm{n,A}}(f) \cdot |H_{\mathrm{n,A}}(f)|^2 \cdot |H_{\mathrm{CMS}}(f)|^2 df, \tag{4.2}$$

where $S_{\mathrm{n,A}}(f)$ refers to the PSD of the current noise source modeling the noise sources at the column-level readout circuitry, and $H_{\mathrm{n,A}}(f)$ represents the transfer function from that source to the output of the column-level amplifier.

As mentioned above, these independent noise sources are uncorrelated. Thus the total input-referred noise charge variance of the readout chain can be expressed as

$$\overline{Q_{\mathrm{n,tot}}^2} = \overline{Q_{\mathrm{n,pix}}^2} + \overline{Q_{\mathrm{n,A}}^2}. \tag{4.3}$$

4.3 Impact of CMS on Noise

Figure 4.3 recalls the principle of CMS. It consists in averaging M samples of the sense node reset level voltage and M other samples of the voltage level after transferring the accumulated charge from the PPD to the sense node. The difference between the second and first averages corresponds to the signal voltage proportional to the transferred charge. The main practical implementations of CMS are depicted in Figs. 4.4 and 4.5. CMS can be performed in the analog domain using sample and holds as shown in Fig. 4.4. This can be implemented using a passive switched capacitors (SC) network or by using an active SC integrator [7]. The CMS can also be performed in the digital domain using registers [8, 9] as shown in Fig. 4.5. Note that both implementations require sample and hold stages with a sampling period of $2MT_{\mathrm{S}}$, where T_{S} is the duration between samples. T_{S} is set to give the signal enough time to settle between two adjacent samples. The sample and hold process can be mathematically modeled in the time domain by a multiplication by a Dirac trail convoluted with a zero order hold. In time domain, the sampled low-pass-filtered signal can be expressed as

$$V_{\mathrm{SH}}(t) = h(t) * \sum_{n=-\infty}^{+\infty} \delta(t - n \cdot 2MT_{\mathrm{S}}) \cdot V_{\mathrm{LP}}(t), \tag{4.4}$$

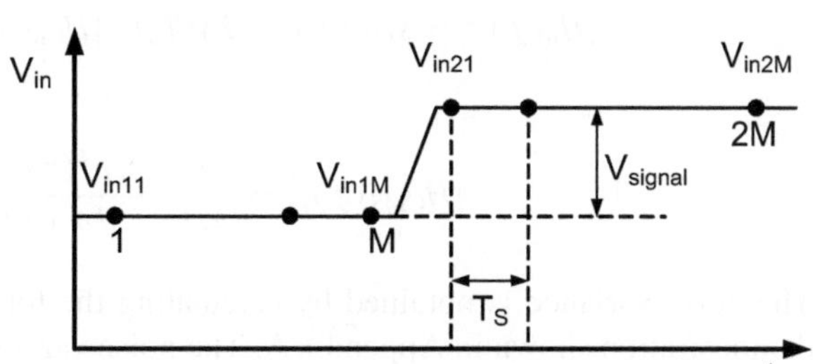

Fig. 4.3 Timing diagram of the an M-order CMS showing the M samples at the first and second voltage levels

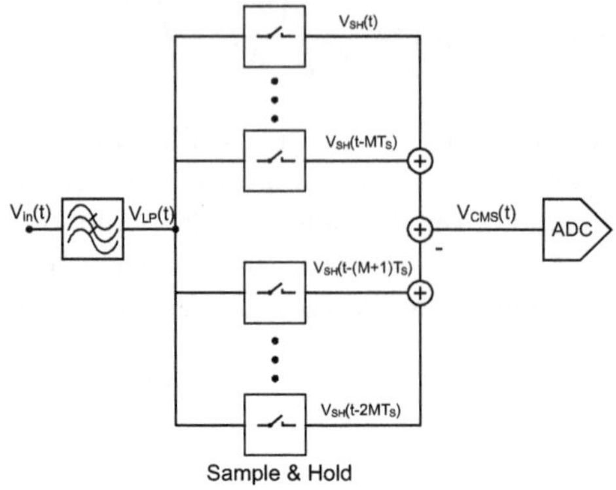

Fig. 4.4 Implementation of the CMS in the analog domain

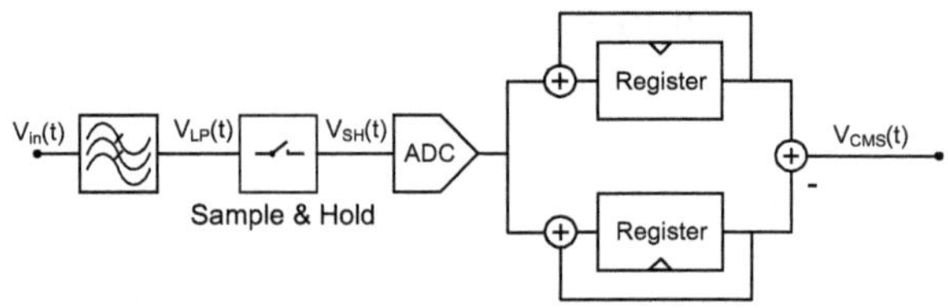

Fig. 4.5 Implementation of the CMS in the digital domain

where $h(t)$ is the rectangular function that takes a unity value in $[0, 2MT_S]$ and zero elsewhere. The signal at the output of the CMS can then be expressed in time domain as

$$V_{CMS}(t) = V_{SH}(t) * \frac{1}{M} \sum_{k=0}^{M-1} \delta(t - kT_S) - \delta(t - (k+M)T_S). \tag{4.5}$$

In order to investigate the impact of correlated sampling on noise, a noise source with a PSD S_n bandlimited at the first order with a cutoff frequency f_c is considered. The noise PSD at the output of the CMS is obtained by transposing (4.5) in the Fourier domain

$$S_{n,CMS}(f) = \sum_{n=-\infty}^{+\infty} |H_n(f)|^2 \cdot S_n(f - \frac{n}{2MT_S}), \tag{4.6}$$

where

$$|H_n(f)|^2 = sinc^2(\pi f \cdot 2MT_S) \cdot |H_{CMS}(f - \frac{n}{2MT_S})|^2 \tag{4.7}$$

and

$$|H_{CMS}(f)|^2 = \frac{4}{M^2} \cdot \frac{sin^4(\pi M f T_S)}{sin^2(\pi f T_S)}. \tag{4.8}$$

The noise variance is obtained by calculating the total noise power. Based on the demonstration shown in Appendix A. The noise variance after performing the CMS can be calculated as

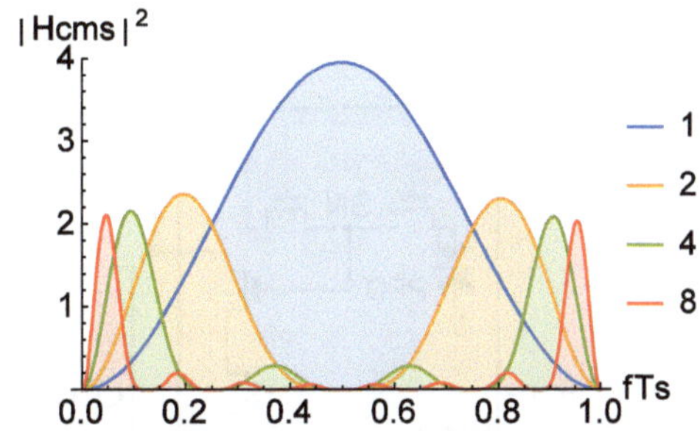

Fig. 4.6 Plot of the squared absolute value of $|H_{\mathrm{CMS}}(f)|^2$

$$\int_{-\infty}^{+\infty} S_{\mathrm{n,CMS}}(f)df = \int_{-\infty}^{+\infty} |H_{\mathrm{CMS}}(f)|^2 \cdot S_{\mathrm{n}}(f)df. \tag{4.9}$$

The term $|H_{\mathrm{CMS}}(f)|^2$ representing the impact of correlated sampling on the noise PSD is plotted in Fig. 4.6 for M ranging between 1 and 8. Notice that the area delimited by $|H_{\mathrm{CMS}}(f)|^2$ is inversely proportional to M, therefore, the white noise variance at the output of the CMS stage is inversely proportional to M. For all values of M, the CMS transfer function applies a zeroing to the low frequencies. The area delimited by $|H_{\mathrm{CMS}}(f)|^2$ reduces by increasing M but the maximum of $|H_{\mathrm{CMS}}(f)|^2$ gets closer to the vertical axis. Thus one can assume that $1/f$ noise is also reduced by increasing the CMS order but this reduction reaches a limit for a certain order M. These aspects are detailed analytically in the following section.

4.4 In-Pixel Source Follower Based Readout Chain

Most of CIS readout chains are based on in-pixel source follower. This scheme offers a simple design, a high dynamic range and simple reset robust against charge injection. It is so far the most used in-pixel readout scheme. Figure 4.7 shows the schematic of a conventional CIS readout chain based on a PPD, an in-pixel source follower stage, column level amplification and correlated sampling. In this section, the signal path is first analyzed in order to derive the detailed expression of the overall conversion gain. Then a noise analysis is performed in order the calculate the expression of the input-referred noise. The noise sources, namely, the thermal noise, the $1/f$ noise and the leakage current shot noise are analyzed separately.

4.4.1 The Conversion Gain

For each noise source, the variance at the output of the readout chain is first calculated and then referred to the input as a noise charge. Hence the pixel conversion gain is

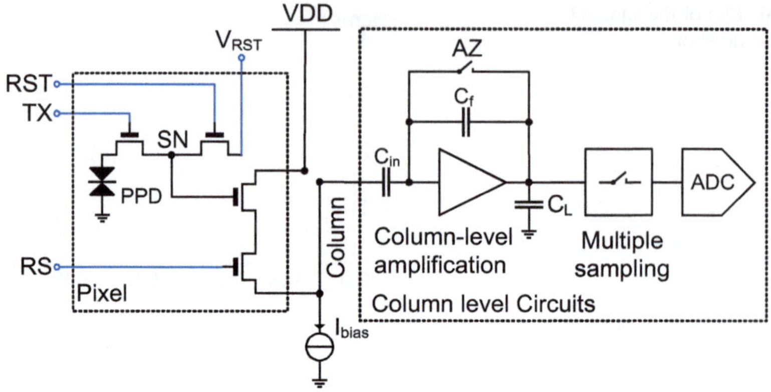

Fig. 4.7 Conventional source follower based CIS readout chain

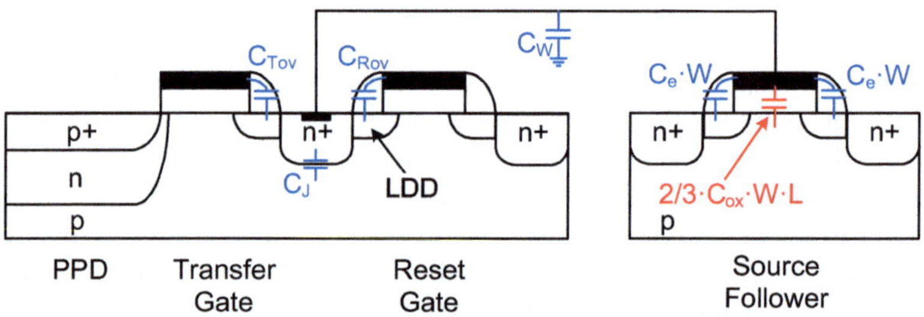

Fig. 4.8 Cross-section of a conventional 4T pixel showing the different parasitic elements contributing to the sense node capacitance

a key parameter in the noise analysis. It is crucial to take into account the effect of parasitic capacitances connected to the sense node. Indeed, these are the first elements in the readout chain. Figure 4.8 presents a schematic of a 4T pixel section view showing all the parasitic capacitances connected to the sense node. These include the overlap capacitances of the transfer and reset gates C_{Tov} and C_{Rov}, respectively, the sense node junction capacitance C_{J} and the parasitic capacitance related to the metal wires C_{W}. These capacitances are independent from the in-pixel SF. Their sum is defined as

$$C_{\mathrm{P}} = C_{\mathrm{Tov}} + C_{\mathrm{Rov}} + C_{\mathrm{J}} + C_{\mathrm{W}}. \tag{4.10}$$

The pixel conversion gain can be calculated using a small-signal analysis of the pixel. The pinned photo diode is modeled with a current source $I_{\mathrm{in}}(t)$ injecting a charge $Q_{\mathrm{in}} = I_{\mathrm{in}} \cdot \Delta t$ in the sense node. The transfer function of the pixel giving the expression of the column level voltage induced by a charge injected in the sense node can be expressed using the simplified small-signal circuit shown in Fig. 4.9. Assuming that $g_{\mathrm{out,SF}} \ll G_{\mathrm{m,SF}}$, where $g_{\mathrm{out,SF}}$ is the output conductance of the source follower and $G_{\mathrm{m,SF}}$ its gate transconductance. The pixel transfer function is given by

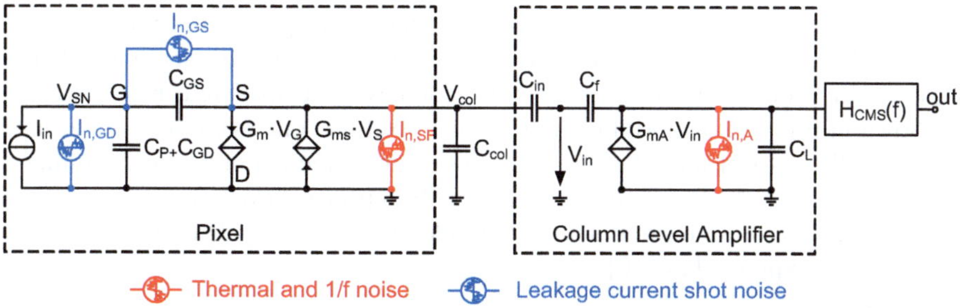

Fig. 4.9 Small-signal analysis of the CIS readout chain depicted in Fig. 4.7 showing the different readout noise sources considered in the analysis

$$H_{\mathrm{pix}}(f) = \frac{V_{\mathrm{col}}}{Q_{\mathrm{in}}} = \frac{A_{\mathrm{CG}}}{1 + j\frac{f}{f_{\mathrm{c,pix}}}}, \tag{4.11}$$

where $Q_{\mathrm{in}} = I_{\mathrm{in}}/j2\pi f$, $f_{\mathrm{c,pix}} = \frac{1}{2\pi}\frac{G_{\mathrm{m,SF}}}{C_{\mathrm{col}}\cdot A_{\mathrm{CG}}\cdot(C_{\mathrm{GS}}+C_{\mathrm{GD}}+C_{\mathrm{P}})}$ is the cut-off frequency of the in-pixel source follower stage, C_{col} is the column level capacitance, C_{GS} and C_{GD} are the gate-to-source and gate-to-drain source follower capacitances, C_{P} is the sum of parasitic capacitances due to wiring, transfer gate, reset transistor and the n^+ junction, and n is the slope factor of the source follower transistor. In saturation, $n = \frac{G_{\mathrm{ms,SF}}}{G_{\mathrm{m,SF}}}$ where $G_{\mathrm{ms,SF}}$ is the source transconductance of the source follower transistor. The value of n ranges from 1.2 to 1.6 and slowly tends to 1 for high V_{G} [10]. The conversion gain corresponds to the dc gain of the pixel transfer function $H_{\mathrm{pix}}(f)$. It is given by

$$A_{\mathrm{CG}} = \frac{\frac{1}{n}}{C_{\mathrm{P}} + C_{\mathrm{GD}} + (1 - \frac{1}{n})C_{\mathrm{GS}}}, \tag{4.12}$$

This expression can be further detailed, by expressing the capacitances C_{GD} and C_{GS} as functions of the source follower gate size and oxide capacitance density per unit area, as

$$A_{\mathrm{CG}} = \frac{\frac{1}{n}}{C_{\mathrm{P}} + C_{\mathrm{e}} \cdot W + (1 - \frac{1}{n})(C_{\mathrm{e}} \cdot W + \frac{2}{3}C_{\mathrm{ox}} \cdot W \cdot L)}, \tag{4.13}$$

where C_{e} is the extrinsic capacitance per unit width of the in-pixel source follower transistor. It includes the overlap and fringing capacitances as depicted in Fig. 4.8. C_{ox} is the SF oxide capacitance per unit area.

4.4.2 Thermal Noise

The column level amplifier has two main roles. The first is the bandwidth control limiting the thermal noise originating from the in-pixel source follower stage that has a large bandwidth. The second is the introduction of a gain high enough to make the contribution of the next stages to the total input-referred noise negligible. Under these assumptions only two thermal noise sources are considered in this analysis as shown in Fig. 4.9: one originating from the pixel and one from the column-level amplifier. We also assume that the open loop gain of the column-level amplifier is provided by an OTA. The thermal noise of a MOS transistor operating in saturation is modeled by a drain current source that adds to the signal. The in-pixel current noise PSD is expressed as [10]

$$S_{I_D}(f) = 4 \cdot k \cdot T \cdot \gamma_{SF} \cdot G_{m,SF}, \tag{4.14}$$

where $G_{m,SF}$ is the gate transconductance of the in-pixel SF transistor, γ_{SF} is the excess noise factor given by $\frac{2n}{3}$, for a long-channel transistor biased in strong inversion [10]. Note than γ_{SF} can also include terms representing the noise contribution of the current mirror used to bias the SF transistor.

Based on the small-signal circuit depicted in Fig. 4.9, the transfer function of the noise originating from the in-pixel source follower referred to the column level is given by

$$H_{n,pix}(f) = \frac{\frac{1}{G_{m,SF}} A_{CG} \cdot (C_{GS} + C_{GD} + C_P)}{1 + j\frac{f}{f_{c,pix}}}. \tag{4.15}$$

The column level amplifier operates in two phases, auto-zeroing and amplification. After auto-zeroing, the noise frozen in the integrating capacitor C_{in} is transferred to the feedback capacitor and to the output during the amplification phase. This frozen noise is canceled thanks to the action of the correlated double sampling at the output of the column amplifier. Thus, for noise calculation, we only need to consider the direct noise at the output of the column amplifier during the amplification phase. The transfer function of the column amplifier when the auto-zeroing switch is opened is calculated based on the small-signal circuit shown in Fig. 4.9. The column level gain is given by $A_{col} = \frac{C_{in}}{C_f}$, where C_{in} and C_f are respectively the integrating and feedback capacitors of the column level amplifier. In order to simplify the noise calculations, it is assumed that $A_{col} \ll \frac{G_{m,A}}{g_{out,A}}$, where $g_{out,A}$ and $G_{m,A}$ are the output conductance and the transconductance of the column level OTA. The zero of the column amplifier transfer function numerator is given by $\frac{1}{2\pi}\frac{A_{col} \cdot G_{m,A}}{C_{in}}$ which is much higher than the cutoff frequency as shown here after by (4.16). Hence its impact has been neglected. The column amplifier transfer function is then expressed as

$$H_A(f) = \frac{-A_{col}}{1 + j\frac{f}{f_{c,A}}}, \tag{4.16}$$

where $f_{c,A} = \frac{1}{2\pi}\frac{G_{m,A}}{(A_{col}+1)C_L + C_{in}}$, C_{in} and C_L are the column amplifier integrating and load capacitors.

At this point, the input-referred thermal noise originating from the pixel can be calculated using (4.1). Similarly, the current noise PSD of the column-level amplifier OTA is expressed as

$$S_{I_D}(f) = 4 \cdot k \cdot T \cdot \gamma_A \cdot G_{m,A}, \tag{4.17}$$

where $G_{m,A}$ is the equivalent transconductance of the operational transconductance amplifier (OTA). γ_A is the excess noise factor of the OTA that takes into account the contribution of all the saturated transistors of the OTA. Note that γ_A increases with the number of saturated transistors of the OTA.

Using the small-signal circuit of Fig. 4.9, the transfer function for the noise originating from the column level amplifier and referred to its output is given by

$$H_{n,A}(f) = \frac{\frac{A_{col}+1}{G_{m,A}}}{1 + j\frac{f}{f_{c,A}}}. \tag{4.18}$$

At this point, the input-referred thermal noise contribution of the column-level amplifier can be calculated using (4.2). The two dominant pixel and column level noise sources are uncorrelated, thus their noise PSDs add. Hence, the total input-referred noise charge variance can be calculated using (4.3) as

$$\overline{Q_{th}^2} = \alpha_{th} \cdot \frac{kT}{A_{col} \cdot C} \left(\frac{\gamma_{SF} G_{m,A} (C_P + C_{GD} + C_{GS})^2}{G_{m,SF}} + \frac{\gamma_A}{A_{CG}^2} \right), \tag{4.19}$$

where $C = C_L + \frac{C_{in}}{A_{col}+1}$. α_{th} is a unitless circuit design parameter reflecting the impact of CMS. α_{th} is given by

$$\alpha_{th} = \frac{1}{\pi f_c} \int_0^\infty \frac{4}{M^2} \frac{sin^4(\pi \cdot M \cdot T_S \cdot f)}{sin^2(\pi \cdot T_S \cdot f)} \cdot \frac{1}{1 + \left(\frac{f}{f_c}\right)^2} df \simeq \frac{2}{M}. \tag{4.20}$$

Figure 4.10 shows α_{th}, normalized to $\frac{2}{M}$ as a function of $f_c \cdot T_S$ (f_c is the cutoff frequency of the readout chain and T_S the CMS sampling period). Note that for proper settling of the signal between sampling instants, $2\pi \cdot f_c \cdot T_S$ has to be typically larger than 5 and under such conditions α_{th} can simply be approximated by $\frac{2}{M}$. This result can be interpreted otherwise without having to go through the calculation. The CMS consists first in averaging M uncorrelated samples (thermal noise). Hence averaging the M uncorrelated samples would result in decreasing the noise variance by $1/M$. Differentiating the two uncorrelated averages of thermal noise would double the noise variance. Thus a CMS of order M results in multiplying the thermal noise variance by a factor of $\frac{2}{M}$. Moreover, the expression of the input-referred thermal noise can be further detailed by expressing the source follower capacitances as well as the transconductance as function of the gate width W and length L. Indeed, the gate to source capacitance can be expressed when the transistor is saturated as the sum of the oxide intrinsic capacitance of the transistor and the extrinsic capacitance

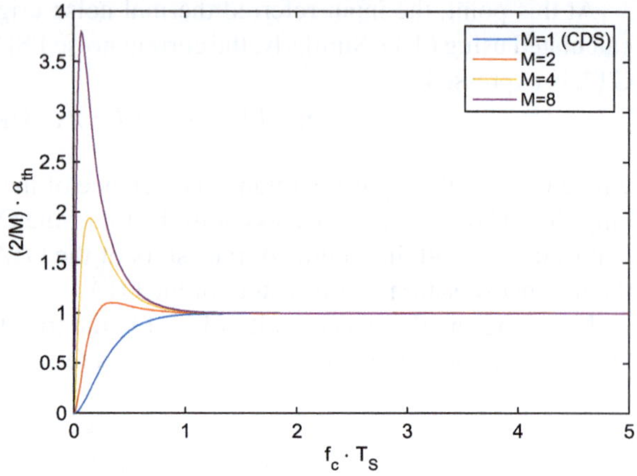

Fig. 4.10 Numerical calculation of the parameter α_{th} from (4.20) as a function of $f_c \cdot T_S$ for a simple CDS and CMS with different orders M

density C_e due to the overlap and fringing:

$$C_{\text{GS}} = \frac{2}{3} \cdot C_{\text{ox}} \cdot W \cdot L + C_e \cdot W, \tag{4.21}$$

while the gate to drain capacitance can be simply expressed as

$$C_{\text{GD}} = C_e \cdot W. \tag{4.22}$$

The transconductance of the in-pixel SF, assuming the transistor is biased in strong inversion and saturation, can be expressed as

$$G_{\text{m,SF}} = \sqrt{\frac{2 \cdot \mu \cdot C_{\text{ox}} \cdot I_D \cdot W}{n \cdot L}}, \tag{4.23}$$

where μ is the charge carrier mobility in the channel of the in-pixel readout transistor and I_D is its bias drain current. The expression of the input-referred thermal noise can then be expressed as

$$\overline{Q_{\text{th}}^2} = \frac{2kT}{M \cdot A_{\text{col}} \cdot C} \left(\frac{\gamma_{\text{SF}} G_{\text{m,A}} (C_{\text{P}} + 2C_e \cdot W + \frac{2}{3} C_{\text{ox}} \cdot W \cdot L)^2}{\sqrt{\frac{2 \cdot \mu \cdot C_{\text{ox}} \cdot I_D \cdot W}{n \cdot L}}} + \frac{\gamma_A}{A_{\text{CG}}^2} \right). \tag{4.24}$$

4.4.3 1/*f* Noise

It is well-known that the drain current $1/f$ noise PSD is inversely proportional to the gate area. In low noise CIS readout chains, the transistors located outside the pixel array and fed with a low impedance signal can be designed with gate dimensions much larger than the in-pixel source follower transistor. In this case, the latter becomes the dominant $1/f$ noise source in the readout chain and the other $1/f$ noise sources can be neglected. Consequently, in the $1/f$ noise analysis, we only consider the noise originating from the pixel.

Based on the discussion of the previous Chapter, the $1/f$ noise current PSD of the source follower can be expressed as

$$S_{I_D}(f) = \frac{K_F}{C_{ox}^2 \cdot W \cdot L \cdot f} \cdot G_{m,SF}^2. \tag{4.25}$$

Using the same noise and signal transfer functions used in the thermal noise analysis based on the small-signal circuit of Fig. 4.9, an analytical expression of the input-referred $1/f$ noise can be obtained despite the divergence of the PSD at $f = 0$. This is due to the impact of the multiple sampling scheme implemented with the CMS/CDS. Indeed, the CMS/CDS transfer function brings the $1/f$ term to 0 at $f = 0$. The input-referred $1/f$ noise can be expressed using (4.1), (4.8) and (4.9) as

$$\overline{Q_{1/f}^2} = \alpha_{1/f} \cdot \frac{K_F(C_P + C_{GD} + C_{GS})^2}{C_{ox}^2 \cdot W \cdot L}, \tag{4.26}$$

where $\alpha_{1/f}$ is a unitless circuit design parameter reflecting the impact of the CMS noise reduction on the $1/f$ noise. Based on the detailed analytical calculation [11], it can be expressed as

$$\alpha_{1/f} = \int_0^{\infty} \frac{1}{f} \cdot \frac{4}{M^2} \frac{sin^4(\pi \cdot M \cdot T_S \cdot f)}{sin^2(\pi \cdot T_S \cdot f)} \cdot \frac{1}{1 + \left(\frac{f}{f_c}\right)^2} df, \tag{4.27}$$

where f_c is the cutoff frequency of the column level amplifier, which is assumed to be lower than the SF stage bandwidth. $\alpha_{1/f}$ is calculated numerically and plotted as a function of $f_c.T_S$ in Fig. 4.11. It shows that $\alpha_{1/f}$ is weakly dependent on T_S when M is higher than 2. In this case $\alpha_{1/f}$ ranges between 3 and 4.

By expressing the capacitances C_{GS} and C_{GD} as a function of the SF gate size and oxide capacitance density per unit area, (4.26) can be further be developed to

$$\overline{Q_{1/f}^2} = \alpha_{1/f} \cdot \frac{K_F(C_P + 2C_e \cdot W + \frac{2}{3}C_{ox} \cdot W \cdot L)^2}{C_{ox}^2 \cdot W \cdot L}. \tag{4.28}$$

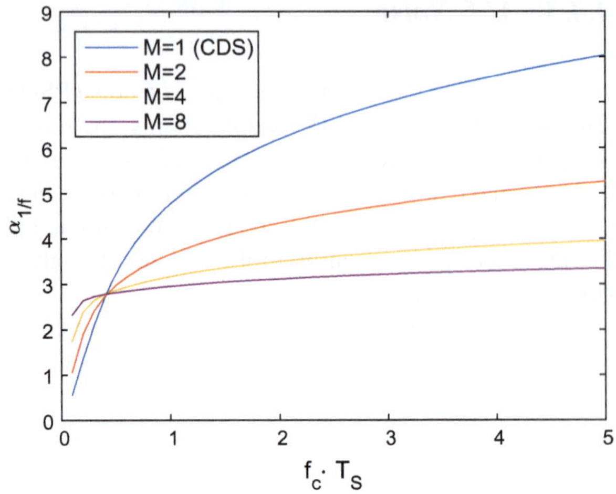

Fig. 4.11 Numerical calculation of the parameter $\alpha_{1/f}$ from (4.27) as a function of $f_c \cdot T_S$ for a simple CDS and CMS with different orders M

4.4.4 Leakage Current Shot Noise

During the readout, the charge transferred to the SN may be corrupted by all the leakage currents through the junctions and gate oxide due to tunneling. Since these leakage currents are due to barrier control processes, they give rise to shot noise. As shown in the small-signal schematic of Fig. 4.9, the leakage current shot noise can be modeled by two noise current sources: $I_{n,GD}$ and $I_{n,GS}$. $I_{n,GD}$ represents the shot noise of all the leakage currents flowing between the SN and the ground, which includes the SN junction leakage and the SF gate oxide tunneling current that sinks into the bulk and the drain. $I_{n,GS}$ represents the shot noise associated to part of the SF gate oxide tunneling current that flows to the source. The unilateral PSD of the current shot noise can be expressed as [12]

$$S_{I_L}(f) = 2 \cdot q \cdot I_L, \tag{4.29}$$

where I_L is the mean value (DC current) of the leakage current. Both shot noise components $I_{n,GD}$ and $I_{n,GS}$ have the same transfer function magnitude, between the noise current source and the output of the column level amplifier. Indeed, the $I_{n,GS}$ source can be represented by two opposite current sources, one parallel to the gate to drain source ($I_{n,GD}$) and the other one parallel to the source to drain thermal noise source. The thermal noise source current PSD is given by $4 \cdot k \cdot T \cdot \gamma_{SF} \cdot G_{m,SF}$ [10]. For $G_{m,SF}$ =10 mS, γ_{SF} =1, the term $2\frac{kT}{q}\gamma_{SF} \cdot G_{m,SF}$ is in the order of 240 µA. The leakage current is of the order of a few tens of fA in 90 nm technologies and a few hundreds of fA for 65 nm technologies. Hence, the source to drain leakage current source effect on the noise is totally negligible compared to the thermal noise of the SF. The pixel leakage current shot noise transfer function can hence be expressed as

$$H_{\mathrm{n,pix}}(f) = \frac{A_{\mathrm{CG}}^2}{(2\pi f)^2} \cdot \frac{1}{1 + \left(\frac{f}{f_{\mathrm{c,pix}}}\right)^2}. \tag{4.30}$$

Note that I_{L} is the sum of all the sense node leakage currents.

$$S_{\mathrm{L,out}}(f) = S_{I_{\mathrm{L}}}(f) \cdot |H_{\mathrm{n,pix}}(f)|^2 \cdot |H_{\mathrm{A}}(f)|^2 \cdot |H_{\mathrm{CMS}}(f)|^2. \tag{4.31}$$

Figure 4.12 shows a plot of the input-referred shot noise PSD, normalized to $2 \cdot q \cdot I_L \cdot T_{\mathrm{S}}$. It can be noticed that due to the $1/f^2$ term in (4.30), the PSD is independent of f_{c} and the area of the PSD increases with M.

Using (4.1), the input-referred charge variance due to the total leakage currents shot noise can be expressed as

$$\overline{Q_{\mathrm{L}}^2} = 2 \cdot \alpha_{\mathrm{shot}} \cdot q \cdot I_{\mathrm{L}} \cdot T_{\mathrm{S}}, \tag{4.32}$$

with

$$\alpha_{\mathrm{shot}} = \frac{1}{T_{\mathrm{S}}} \cdot \int_0^{\infty} \frac{1}{(2\pi f)^2} \cdot \frac{4}{M^2} \frac{sin^4(\pi \cdot M \cdot T_{\mathrm{S}} \cdot f)}{sin^2(\pi \cdot T_{\mathrm{S}} \cdot f)} \cdot \frac{1}{1 + \left(\frac{f}{f_{\mathrm{c}}}\right)^2} df. \tag{4.33}$$

Figure 4.13 shows a numerical evaluation of α_{shot} as a function of the CMS order M. It shows that:

$$\alpha_{\mathrm{shot}} \simeq \begin{cases} \frac{M}{3} \text{ for } M \geq 2 \\ \frac{1}{2} \text{ for } M = 1 \end{cases} \tag{4.34}$$

Due to the $1/f^2$ term, the impact of the low pass filtering is completely negligible and the value of α_{shot} is independent of the cutoff frequency of the readout chain. Note that the shot noise current sources feature a white PSD. But when integrated in the SN capacitance, they give rise to a Wiener process [13]. The variance of this noise is thus expected to rise with the readout time. In order to evaluate the impact

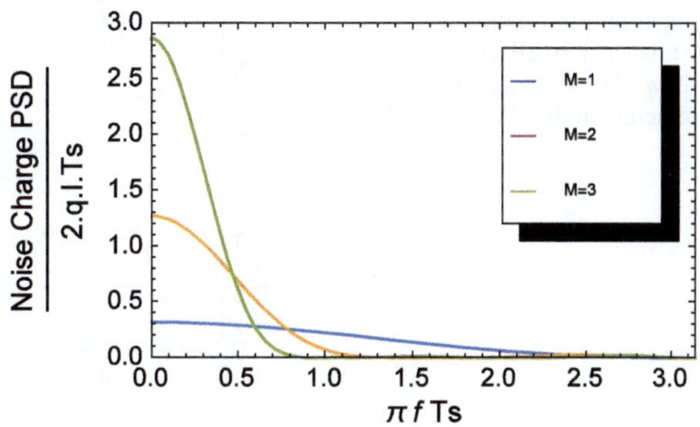

Fig. 4.12 Numerical evaluation of the input-referred shot noise PSD normalized to $2 \cdot q \cdot I_L \cdot T_{\mathrm{S}}$ as a function of the CMS order M

of the CMS on the leakage current shot noise, α_{shot} is calculated numerically and plotted in Fig. 4.13 as a function of M. In the case of a simple CDS α_{shot} is equal to 0.5, hence, the shot noise variance is given by $q \cdot I_L \cdot T_S$ which corresponds to a typical case of a Wiener process [13]. Figure 4.13 also shows that, in the general case, the leakage current shot noise increases linearly with $T_S \cdot M$.

4.5 In-Pixel Common Source Based Readout Chain

One other possible in-pixel readout scheme consists in operating the in-pixel amplifying transistor in the common source configuration. This configuration results in a voltage amplification with a capacitive feedback. Indeed, this capacitive feedback is at least due to the gate to drain parasitic capacitance of the common source transistor. Hence, this configuration corresponds to an in-pixel capacitive trans-impedance amplifier (CTIA). This solution presents a voltage gain at the earliest stage of the readout chain. Figure 4.14 shows the schematic of a conventional CIS readout chain based on a PPD, an in-pixel common source stage, column level amplification and correlated sampling. In this section, the signal path is first analyzed in order to derive the detailed expression of the overall conversion gain. Then a noise analysis is performed in order the calculate the expression of the input-referred noise. The noise sources, namely, the thermal noise, the $1/f$ noise and the leakage current shot noise are analyzed separately.

4.5.1 *Conversion Gain*

As discussed in the previous section, it is very important to perform a detailed calculation of the conversion gain taking into account the parasitic capacitances connected

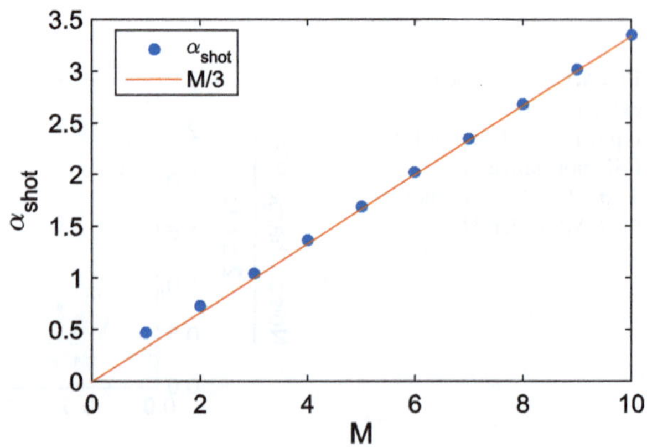

Fig. 4.13 Numerical evaluation of the input-referred shot noise variance normalized to $2 \cdot q \cdot I_L \cdot T_S$ (α_{shot}) as a function of the CMS order M

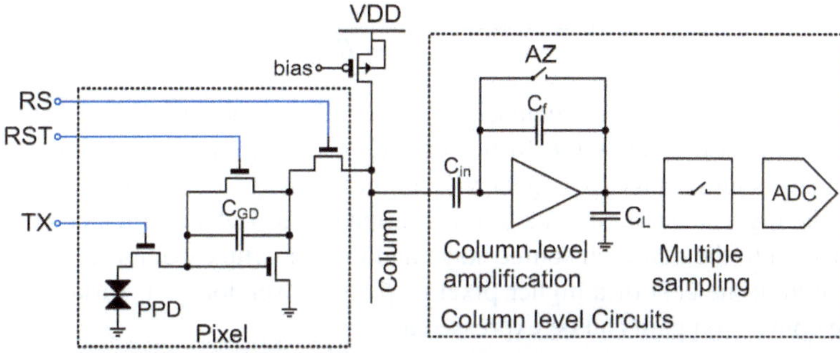

Fig. 4.14 In-pixel Common-source (CTIA) based CIS readout chain

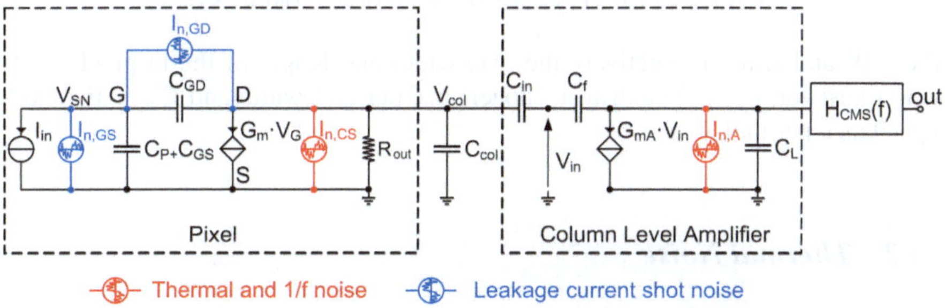

Fig. 4.15 Small-signal analysis of the CIS readout chain depicted in Fig. 4.14 showing the different readout noise sources considered in the analysis

to the sense node. Based on Fig. 4.8 depicting the different capacitances involved in the charge to voltage conversion and small signal circuit shown in Fig. 4.15, the transfer function relating the input charge pulse (modeling the charge injection from the PPD into the sense node) to the output voltage of the pixel (column voltage) can be expressed as

$$H_{\text{pix}}(f) = \frac{V_{\text{col}}}{Q_{\text{in}}} = \frac{A_{\text{CG}}}{1 + j\frac{f}{f_{\text{c,pix}}}}, \tag{4.35}$$

where $f_{\text{c,pix}}$ is given by $\frac{1}{2\pi R_{out} C_{\text{col}}}$. C_{col} is the load capacitance of the common source stage mainly given by the column total parasitic capacitance. R_{out} is the output resistance of the common source stage given by the parallel output resistances of the in-pixel common source transistor and its load. Note that the load of the common source stage can be implemented by a passive resistance or by a current mirror. But in the latter case, the output resistance might be very large, which would result in a low bandwidth.

The pixel conversion gain A_{CG} is given by

$$A_{\mathrm{CG,CS}} = \frac{R_{out} \cdot G_{\mathrm{m,CS}}}{C_{\mathrm{P}} + C_{\mathrm{GS}} + C_{\mathrm{GD}} + R_{out} \cdot G_{\mathrm{m,CS}} \cdot C_{\mathrm{GD}}}. \tag{4.36}$$

where $G_{\mathrm{m,CS}}$ is the transconductance of the common source transistor. The term $R_{out} \cdot G_{\mathrm{m,CS}} \cdot C_{\mathrm{GD}}$ reflects the Miller effect. $R_{out} \cdot G_{\mathrm{m,CS}}$ is the voltage gain of the common source stage, it is generally higher than 10. Equation (4.36) suggests that for the highest conversion gain, no additional feedback capacitance need to be implemented to the common source stage and the parasitic capacitance of the latter can be used at the cost of a higher pixel-to-pixel conversion gain variation. In this case the conversion gain can be expressed as

$$A_{\mathrm{CG,CS}} = \frac{R_{out} \cdot G_{\mathrm{m,CS}}}{C_{\mathrm{P}} + 2C_{\mathrm{e}} \cdot W + \frac{2}{3}C_{\mathrm{ox}} \cdot W \cdot L + R_{out} \cdot G_{\mathrm{m,CS}} \cdot C_{\mathrm{e}} \cdot W}, \tag{4.37}$$

where W and L are respectively the gate width and length of the in-pixel amplifying transistor, C_{e} is the extrinsic capacitance per unit width and C_{ox} is the oxide capacitance per unit area.

4.5.2 *Thermal Noise*

The CTIA based readout chain presents a voltage gain and bandwidth limitation at the pixel-level, which makes the thermal noise performance less dependent on the column-level amplifier compared to the source follower based readout chain. Nevertheless, the latter remain an important block. Indeed, typical conversion gains of pixels implementing the in-pixel common source stage are of the order of a few $100\,\mu\mathrm{V/e^-}$ [14] which is still not enough to make the noise contribution of the next stages (buffers, CMS, ADC) totally negligible. Hence, for the ultimate noise performance, column-level amplification is still required. The column level amplifier signal and noise transfer functions are calculated in the previous section. The analysis of the thermal noise originating from the column-level amplifier remain the same. For the thermal noise originating from the pixel, the small signal circuit shown in Fig. 4.15 is used. The in-pixel current noise PSD is expressed as [10]

$$S_{I_{\mathrm{D}}}(f) = 4 \cdot k \cdot T \cdot \gamma_{\mathrm{CS}} \cdot G_{\mathrm{m,CS}}, \tag{4.38}$$

where γ_{CS} reflects the excess noise factor of the common source stage. The noise transfer function relating the drain noise current source to the column-level voltage can be expressed as

$$H_{\mathrm{n,pix}}(f) = A_{\mathrm{CG,CS}} \cdot \frac{C_{\mathrm{P}} + C_{\mathrm{GD}} + C_{\mathrm{GS}}}{G_{\mathrm{m,CS}}} \cdot \frac{1}{1 + j\frac{f}{f_{\mathrm{c,pix}}}}. \tag{4.39}$$

In the source follower based configuration, the pixel noise transfer function features a cutoff frequency of $\frac{G_{m,SF}}{C_{col}}$ whereas it is given by $\frac{1}{R_{out} \cdot C_{col}}$ for the CS amplifier. Generally, $G_{m,SF}$ is much higher than $\frac{1}{R_{out}}$. Hence, the bandwidth control can be performed at the pixel level for this configuration while it requires additional load capacitance of several pF in the case of the source follower configuration.

Based on (4.16), (4.18), (4.36) and (4.39) the total input-referred thermal noise can be expressed using (4.3) as

$$\overline{Q_{th}^2} = \alpha_{th} \cdot k \cdot T \cdot \left(\frac{(C_P + C_{GS} + C_{GD})^2 \cdot \gamma_{CS}}{R_{out} \cdot G_{m,CS} \cdot C_{col} + A_{col} \cdot \frac{G_{m,CS}}{G_{m,A}} \cdot C} + \frac{\gamma_A}{A_{col} \cdot A_{CG}^2 \cdot C} \right). \quad (4.40)$$

This expression can be further detailed by expressing the parasitic capacitances of the in-pixel amplifying transistor as functions of the its gate width, length and capacitance densities

$$\overline{Q_{th}^2} = \alpha_{th} \cdot k \cdot T \cdot \left(\frac{(C_P + 2C_e \cdot W + \frac{2}{3} C_{ox} \cdot W \cdot L)^2 \cdot \gamma_{CS}}{R_{out} \cdot G_{m,CS} \cdot C_{col} + A_{col} \cdot \frac{G_{m,CS}}{G_{m,A}} \cdot C} + \frac{\gamma_A}{A_{col} \cdot A_{CG}^2 \cdot C} \right). \quad (4.41)$$

This expression features two terms, the first one represents the thermal noise originating from the pixel and the second one represents contribution of the column level amplifier. The thermal noise originating from the pixel is, for a given readout chain bandwidth, lower than for the source follower based readout chain. Indeed, in the common source configuration, the bandwidth is controlled at both pixel and column levels. Thus, a second order low pass filtering is performed leading to less thermal noise for the same bandwidth compared to the in-pixel source follower based configuration. The thermal noise contribution of the column-level amplifier is also more reduced in the common source based readout chain since the conversion gain is generally higher.

4.5.3 1/*f* Noise

For the $1/f$ noise analysis in the common source based configuration, the noisy transistors outside the pixel array and fed with a low impedance signal are considered to have gate sizes much larger than the in-pixel ones, thus their $1/f$ noise contribution can be neglected. In the small signal circuit of Fig. 4.15 the $1/f$ noise current PSD is expressed as [10]

$$S_{I_D}(f) = \frac{K_F}{C_{ox}^2 \cdot W \cdot L \cdot f} \cdot G_{m,CS}^2. \quad (4.42)$$

The small-signal circuit of Fig. 4.15 and (4.1) lead to the same input-referred $1/f$ noise expression as the source follower configuration

$$\overline{Q_{1/f}^2} = \alpha_{1/f} \cdot \frac{K_F(C_P + C_{GD} + C_{GS})^2}{C_{ox}^2 \cdot W \cdot L}. \tag{4.43}$$

The best input referred noise is obtained for the lowest values of the capacitances connected to the sense node, namely C_P, C_{GS} and C_{GD}. One can notice that any additional capacitance to the gate to drain parasitic feedback capacitance of the in-pixel common source would simply result in a $1/f$ noise increase. Hence, the optimal way to implement an in-pixel CTIA is by adding no feedback capacitance to the common source stage, in which case, the input-referred $1/f$ noise would be exactly the same as the one calculated in the case of in-pixel SF stage.

$$\overline{Q_{1/f}^2} = \alpha_{1/f} \cdot \frac{K_F(C_P + 2C_e \cdot W + \frac{2}{3}C_{ox} \cdot W \cdot L)^2}{C_{ox}^2 \cdot W \cdot L}, \tag{4.44}$$

4.5.4 Leakage Current Shot Noise

As for the $1/f$ noise, both CIS readout chains based on source follower and common source configurations have the same input referred leakage current shot noise expressed as

$$\overline{Q_L^2} = 2 \cdot \alpha_{shot} \cdot q \cdot I_L \cdot T_S. \tag{4.45}$$

Even if the CS configuration has a second order low pass filtering, the latter has no impact on the value of α_{shot} due to the $1/f^2$.

4.6 Discussion

From a practical aspect, the source follower based readout chain presents several key advantages compared to the common source based one. Indeed, the source follower based pixel can be reset at a wide range of voltages whereas the common source stage requires a feedback reset in order to set the input voltage (sense node) in the linear region of the CS amplifier and avoid the saturation region. Another problem is that the linear part of the common source stage corresponds to low input voltages especially for a an NMOS based common source stage. This low reset voltage might insufficient for an efficient charge transfer from the PPD. In addition the source follower based pixel allows to exploit a higher dynamic range at the pixel output compared to the CTIA where the pixel-level gain limits the dynamic range at the early beginning of the readout chain. Moreover, by comparing the conversion gain expressions of the source follower (4.12) and common source based pixels (4.36), it appears clearly that the conversion gain of the source follower based pixel is independent of the in-pixel amplifying transistor G_m while for the common source based pixel, it is proportional

Table 4.1 Noise analysis summary comparing CIS readout chains sharing the same CMS and column-level amplification and based respectively on in-pixel common source and source follower configuration

Pixel configuration:	SF	CS
Pixel conversion gain	$\dfrac{\frac{1}{n}}{C_P + C_e \cdot W + (1 - \frac{1}{n})(C_e \cdot W + \frac{2}{3} C_{ox} \cdot W \cdot L)}$	$\dfrac{R_{out} \cdot G_{m,CS}}{C_P + 2C_e \cdot W + \frac{2}{3} C_{ox} \cdot W \cdot L + R_{out} \cdot G_{m,CS} \cdot C_e \cdot W}$
Input-referred $1/f$ noise	$\alpha_{1/f} \cdot \dfrac{K_F(C_P + 2C_e \cdot W + \frac{2}{3} C_{ox} \cdot W \cdot L)^2}{C_{ox}^2 \cdot W \cdot L}$	
Input-referred thermal noise originating from the pixel	$\alpha_{th} \cdot \dfrac{kT}{A_{col} \cdot C} \left(\dfrac{\gamma_{SF} G_{m,A} (C_P + 2C_e \cdot W + \frac{2}{3} C_{ox} \cdot W \cdot L)^2}{G_{m,SF}} \right)$	$\alpha_{th} \cdot kT \cdot \left(\dfrac{(C_P + 2C_e \cdot W + \frac{2}{3} C_{ox} \cdot W \cdot L)^2 \cdot \gamma_{CS}}{R_{out} \cdot G_{m,CS} \cdot C_{col} + A_{col} \cdot \frac{G_{m,CS}}{G_{m,A}} \cdot C} \right)$
Input-referred thermal noise originating from the column-level amplifier	$\alpha_{th} \cdot \dfrac{kT \cdot \gamma_A}{A_{CG}^2 \cdot A_{col} \cdot C}$	
Input-referred leakage current shot noise	$2 \cdot \alpha_{shot} \cdot q \cdot I_L \cdot T_S$	

to the product of the in-pixel amplifying transistor G_m and the load resistance R_{out}. Hence, the conversion gain of the common source based pixel is strongly affected by the process variations leading to a high photo-response non uniformity at the pixels array level. The source follower based pixel, on the other hand, is much more robust against mismatch and process variations.

Based on the noise analysis presented above, the two configurations can now be compared from the noise perspective. Table 4.1 presents a summary of the noise analysis presented in this chapter. It shows that both configurations share the same $1/f$ and leakage current shot noise. The common source configuration presents a lower thermal noise for a given bandwidth thanks to the pixel-level gain further minimizing the thermal noise contribution of the column-level amplifier. This theoretical comparison of the noise performance have been confirmed with transient noise simulation results of optimized readout chains based on in-pixel source followers and in-pixel common source amplification presented in [15].

References

1. C. Enz, G. Temes, Circuit techniques for reducing the effects of op-amp imperfections: autozeroing, correlated double sampling, and chopper stabilization. Proc. IEEE **84**(11), 1584–1614 (1996)
2. E.R. Fossum, Charge transfer noise and lag in cmos active pixel sensors, in *Proceedings of 2003 IEEE Workshop on Charge-Coupled Devices and Advanced Image Sensors* (2003), pp. 1–6
3. K. Thornber, Noise suppression in charge transfer devices. Proc. IEEE **60**(9), 1113–1114 (1972)
4. J.R. Janesick, *Scientific Charge-Coupled Devices* (SPIE, USA, 2001), p. 656
5. A. Boukhayma, A. Peizerat, C. Enz, Temporal readout noise analysis and reduction techniques for low-light cmos image sensors. IEEE Trans. Electron Devices **63**(1), 72–78 (2016)
6. A. Boukhayma, A. Peizerat, C. Enz, A correlated multiple sampling passive switched capacitor circuit for low light cmos image sensors, in *2015 International Conference on Noise and Fluctuations (ICNF)* (2015), pp. 1–4
7. S. Suh, S. Itoh, S. Aoyama, S. Kawahito, Column-parallel correlated multiple sampling circuits for cmos image sensors and their noise reduction effects. Sensors **10**(10), 9139–9154 (2010)
8. Y. Chen, Y. Xu, Y. Chae, A. Mierop, X. Wang, A. Theuwissen, A 0.7e-rms temporal-readout-noise cmos image sensor for low-light-level imaging, in *2012 IEEE International Solid-State Circuits Conference Digest of Technical Papers (ISSCC)* (2012), pp. 384–386
9. Y. Lim, K. Koh, K. Kim, H. Yang, J. Kim, Y. Jeong, S. Lee, H. Lee, S.-H. Lim, Y. Han, J. Kim, J. Yun, S. Ham, Y.-T. Lee, A 1.1 e- temporal noise 1/3.2-inch 8 mpixel cmos image sensor using pseudo-multiple sampling, in *2010 IEEE International Solid-State Circuits Conference Digest of Technical Papers (ISSCC)* (2010), pp. 396–397
10. C. Enz, E. Vittoz, *Charge based MOS Transistor Modeling, The EkV Model for Low-Power and RF IC Design* (Wiley, Hoboken, 2006)
11. C. Enz, A. Boukhayma, Recent trends in low-frequency noise reduction techniques for integrated circuits, in *2015 International Conference on Noise and Fluctuations (ICNF)* (2015), pp. 1–6
12. A. Scholten, L. Tiemeijer, R. van Langevelde, R. Havens, A. Zegers-van Duijnhoven, V. Venezia, Noise modeling for rf cmos circuit simulation. IEEE Trans. Electron Devices **50**(3), 618–632 (2003)
13. A. Papoulis, *Probability, Random Variables, and Stochastic Processes* (McGraw-Hill, New York, 1965), pp. 345–396

14. C. Lotto, P. Seitz, T. Baechler, A sub-electron readout noise cmos image sensor with pixel-level open-loop voltage amplification, in *2011 IEEE International Solid-State Circuits Conference* (2011), pp. 402–404
15. A. Boukhayma, A. Peizerat, A. Dupret, C. Enz, Comparison of two optimized readout chains for low light cis, in *Proceedings of SPIE*, vol. 9022 (2004), pp. 90 220H–90 220H–11. https://doi.org/10.1117/12.2036740

Chapter 5
Noise Reduction in CIS Readout Chains

The readout noise analysis detailed in the previous chapter led to the derivation of the input-referred read noise for CIS readout chains based on in-pixel source follower and common source stage. For the $1/f$ noise, the two configurations are equivalent but for the thermal noise, the common source stage may have a slight advantage. Since the source follower configuration is the most commonly used for advantages like simple implementation, dynamic range flexibility and low pixel-to-pixel non uniformity, this chapter focuses on the readout noise reduction in the context of an in-pixel source follower based CIS readout chain. The analytical formulas (4.24), (4.28) involve process and design parameters at different levels of the readout chain. This chapter exploits these formulas by suggesting optimizations including design choices and process refinements for low input-referred noise. It also shows how designers can take advantage of device choices from standard libraries to deeply reduce the input-referred read noise without having to go through any process-level refinements. The thermal and $1/f$ noise reduction techniques are discussed separately. The input referred noise reduction follows two steps. The first consists in optimizing the design parameters of the blocks located outside the pixel, at the column level where more degrees of freedom are available in terms of area and circuit topology. These optimizations are independent of the in-pixel design. The second step consists in pixel-level optimization where only a few degrees of freedom are left to the designer.

5.1 Thermal Noise Reduction

The input-referred thermal noise expression have been derived in (4.24). The expression is recalled using (4.20) and (4.24) for the sake of clarity

A. Boukhayma, *Ultra Low Noise CMOS Image Sensors*,
Springer Theses, https://doi.org/10.1007/978-3-319-68774-2_5

$$\overline{Q_{\text{th}}^2} = \frac{2kT}{M \cdot A_{\text{col}} \cdot C} \left(\frac{\gamma_{\text{SF}} G_{\text{m,A}} (C_{\text{P}} + 2C_{\text{e}} \cdot W + \frac{2}{3} C_{\text{ox}} \cdot W \cdot L)^2}{\sqrt{\frac{2 \cdot \mu \cdot C_{\text{ox}} \cdot I_{\text{D}} \cdot W}{n \cdot L}}} + \frac{\gamma_{\text{A}}}{A_{\text{CG}}^2} \right). \quad (5.1)$$

This expression involves design and process parameters related to the in-pixel source follower stage, the column-level amplifier and the CMS. The pixel-level and column level blocks optimizations are discussed separately.

5.1.1 Column-Level Circuits Optimization

Based on (5.1), the readout thermal noise can be reduced independently of the pixel design using the following parameters: the column level gain A_{col}, the capacitance C determining the bandwidth of the column level amplifier and the CMS order M.

The column level gain A_{col}, the capacitance C and the CMS order M all have equivalent impacts on the input-referred noise reduction. Indeed, the input-referred thermal noise variance is inversely proportional to their product. In order to validate this result, transient noise simulations [1] have been performed on a conventional CIS readout chain with a 4T pixel using a standard thick oxide NMOS source follower transistor, a column level amplifier based on an operational transconductance amplifier (OTA) and the passive switched-capacitor CMS circuit presented in [2]. Figure 5.1 shows the impact of the column level gain and bandwidth control on the input-referred thermal noise when a simple correlated double sampling is used after the column level amplifier. Figure 5.2 shows the impact of the correlated multiple sampling on the input-referred thermal noise for different column level gains. These simulation results show that the thermal noise can be reduced drastically using only the design parameters of the blocks located at the column level.

It is also important to note that the parameters C, A_{col} and M have an equivalent impact on the readout time and the frame rate. Indeed, as shown previously, the

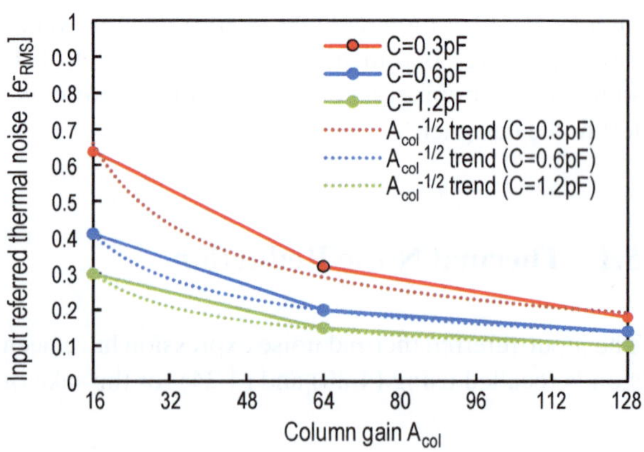

Fig. 5.1 Input-referred thermal noise, obtained from transient noise simulations, of a CIS readout chain, with a 4T pixel (standard NMOS source follower), column amplification (closed loop gain with OTA) and CDS as a function of the column-level gain A_{col} for different values of C

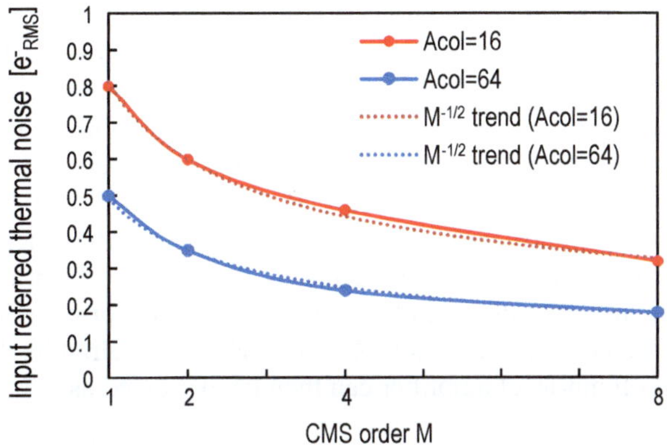

Fig. 5.2 Input-referred thermal noise, obtained from transient noise simulations, of a CIS readout chain, with a 4T pixel (standard NMOS source follower), column amplification (closed loop gain with OTA) and CMS implemented with the analog circuit presented in [2], as a function of the CMS order M for different values of the column level gain A_{col}

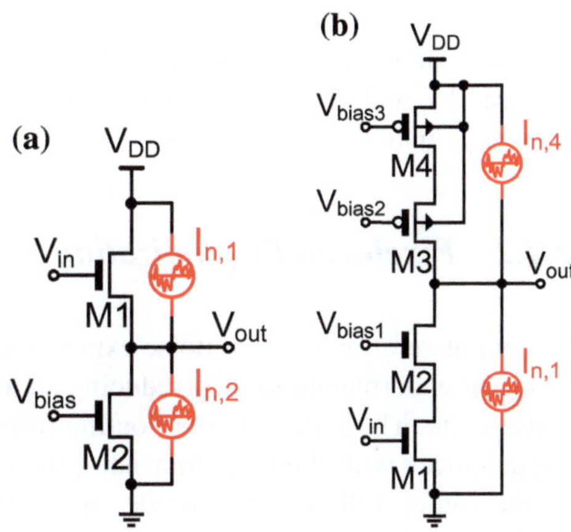

Fig. 5.3 Simplified schematic of a source follower stage with a transistor load (**a**) and a single input fully cascoded amplifier (**b**)

cutoff frequency of the column-level amplifier is inversely proportional to $A_{col} \cdot C$. Performing a CMS of order M requires a readout time M times higher than a simple CDS. Hence, increasing the product $A_{col} \cdot C$ by a factor M while using a CDS will result in the same thermal noise and readout time.

The optimization of the excess noise factors γ_{SF} and γ_A of the source follower stage and column level amplifier, respectively, can also be used to reduce the thermal noise. Figure 5.3 shows the schematics of a source follower stage and a folded cascode amplifier. For CIS, the transistor M_1 in Fig. 5.3a corresponds to the in-pixel readout transistor and M_2 is a transistor load common to the pixels of the column. The noise excess factor of the source follower stage is given by

$$\gamma_{SF} = \gamma_1 + \gamma_2 \frac{G_{m,2}}{G_{m,1}}, \tag{5.2}$$

where γ_1 and γ_2 are the thermal excess noise factors of M_1 and M_2, respectively. $G_{m,1}$ and $G_{m,2}$ are their respective transconductances. Hence, the thermal noise contribution of the transistor M_2 can be minimized by reducing the $\frac{G_{\mathrm{m},2}}{G_{\mathrm{m},1}}$ ratio. This can be achieved for instance by making the aspect ratio of M_2 lower than M_1.

For the column-level amplifier OTA, the noise excess factor can first be optimized by choosing an OTA design with a minimum number of noisy transistors. The basic stage is the common source, but the voltage gain provided by the latter is not enough for closed loop gains of several orders of magnitude. Hence it is necessary to cascode the common source stage. Figure 5.3b shows the schematic of a fully cascoded single input amplifier. The advantage of cascoding is the dramatic increase of the open loop gain without a significant impact on the noise. Indeed the noise contribution of the cascode transistors $M2$ and $M3$ are negligible. The noise excess factor of the column-level amplifier can then be expressed as

$$\gamma_{\mathrm{A}} = \gamma_1 + \gamma_4 \frac{G_{\mathrm{m},4}}{G_{\mathrm{m},1}}. \tag{5.3}$$

Consequently, the excess noise factor of the column level amplifier can be also reduced by making the transconductance $G_{\mathrm{m},4}$ of M_4 lower, as much as possible, than $G_{\mathrm{m},1}$ using the aspect ratio.

5.1.2 *Pixel-Level Optimization*

The input-referred thermal noise expression (5.1) shows that another way to further reduce the thermal noise is by reducing all the parasitic capacitances connected to the sense node. These include the floating diffusion junction capacitance, the coupling capacitances with the reset transistor, the transfer gate and the metal wires as well as the source follower gate oxide and overlap capacitances. Indeed, the reduction of these capacitances increases the charge to voltage conversion gain at the level of the sense node resulting in a higher signal level at the input node of the readout chain. From the designer perspective, the sense node total capacitance reduction can be achieved by using minimum size transistors and careful layout. Figure 5.4 shows a plot of the input-referred thermal noise (5.1) as a function of the in-pixel source follower gate dimensions for a column-level of 64 and a readout chain bandwidth of 250 kHz set by the column-level amplifier. Figure 5.4 shows that a minimum sized in-pixel source follower is optimal for thermal noise reduction.

5.2 Flicker Noise Reduction

The input-referred $1/f$ noise expression has been derived in (4.28). It is recalled for the sake of clarity.

$$\overline{Q_{1/f}^2} = \alpha_{1/f} \cdot \frac{K_{\mathrm{F}}(C_{\mathrm{P}} + 2C_{\mathrm{e}} \cdot W + \frac{2}{3}C_{\mathrm{ox}} \cdot W \cdot L)^2}{C_{\mathrm{ox}}^2 \cdot W \cdot L}. \tag{5.4}$$

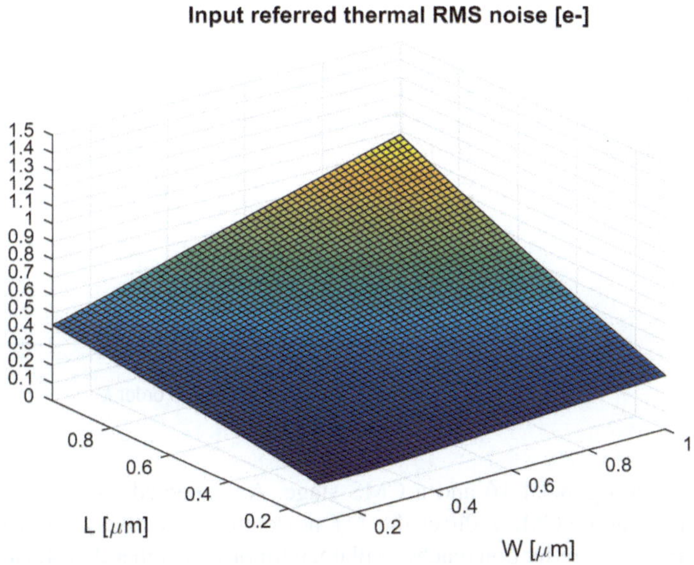

Fig. 5.4 Calculated input-referred thermal noise using (5.1) as a function of the in-pixel source follower transistor gate width W and length L for $\gamma_{\mathrm{SF}} = \gamma_{\mathrm{A}} = 1$, $\sqrt{\frac{2\mu C_{\mathrm{ox}} I_D}{n}} = 12\,\mu\mathrm{S}$, $G_{\mathrm{m,A}} = 30\,\mu\mathrm{S}$ with $A_{\mathrm{col}} = 64$ and $C = 0.3\,\mathrm{pF}$ for a readout chain bandwidth of 250 kHz. The capacitance C_{P} and the source follower capacitances are taken for a 180 nm process as $C_{\mathrm{P}} = 0.75\,\mathrm{fF}$, $C_{\mathrm{ox}} = 4.5\,\mathrm{fF}/\mu\mathrm{m}^2$ and $C_{\mathrm{e}} = 0.45\,\mathrm{fF}/\mu\mathrm{m}$ and the slope factor n of 1.2

This expression is valid when all the readout chain transistors operating in saturation region and located outside the pixel feature a gate area larger enough than the the in-pixel source follower in order to make their contribution to the total $1/f$ noise negligible. This condition is easy to validate since all the blocks located outside the pixel are common to the whole column, which makes their gate area constraint very flexible. The expression (5.4) involves design and process parameters related to the in-pixel source follower stage, the column-level amplifier and the CMS.

5.2.1 Correlated Sampling

The $1/f$ noise reduction starts outside the pixel by using devices with gate areas large enough to make there contribution to the total $1/f$ noise negligible compared to the one originating from the in-pixel amplifying transistor. In this case, $\alpha_{1/f}$ becomes the only parameter in the input-referred $1/f$ noise expression (5.4) which is independent from the pixel-level device and process parameters. As shown theoretically in Fig. 4.11 from the previous chapter, $\alpha_{1/f}$ decreases with the CMS order. Figure 5.5 shows transient noise simulations of the input-referred $1/f$ noise of a readout chains with a 4T pixel based on a thick oxide NMOS source follower, a column level

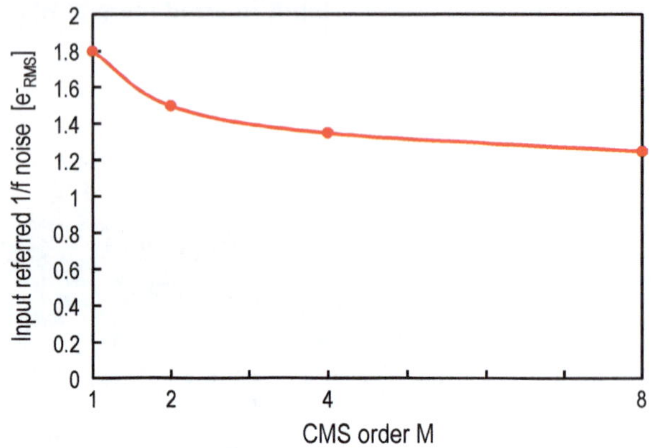

Fig. 5.5 Input-referred $1/f$ noise of a CIS readout chain, with a conventional 4T pixel, column amplification and CMS [2], obtained with transient noise simulations, as a function of the CMS order M

amplifier set to a gain of 16 and a CMS stage. As expected analytically, Fig. 5.5 demonstrates that the CMS reduces the $1/f$ noise more effectively for orders lower that 4. But this noise reduction reaches a plateau for orders higher than 4. Hence, even if the CMS comes with about 20% $1/f$ noise reduction, the impact of the process and device parameters of the in-pixel amplifying transistor is more significant. These aspects will be discussed in the following section.

5.2.2 *Pixel-Level Optimization*

The thermal noise can be reduced to extremely low levels by implementing column level circuit techniques as shown in Figs. 5.1 and 5.2. Figure 5.5 shows that the $1/f$ noise dominates clearly the thermal noise when the latter is minimized using the circuit design parameters described above. Indeed, with a standard thick oxide NMOS source follower, the input-referred $1/f$ noise remains above $1\,e^-_{rms}$. Consequently, the noise optimization at the pixel level should be mostly focused on reducing the remaining and dominant $1/f$ noise. Equation (5.4) is the starting point for the $1/f$ noise optimization. This equation suggests different approaches including proper device selection, design optimizations as well as process improvements.

5.2.2.1 Reduce the $1/f$ Noise Process Parameters

In the input-referred $1/f$ noise expression (5.4), the process dependent parameter directly related to the $1/f$ noise mechanism is K_F which is proportional to the oxide trap density N_t as shown by (3.37). The reduction of K_F can be addressed by making a good device choice in the standard library or through process-level improvements. At the process level, it is known that buried channel devices exhibit a lower $1/f$ noise by featuring a lower K_F parameter. This is likely due to the fact that the charge

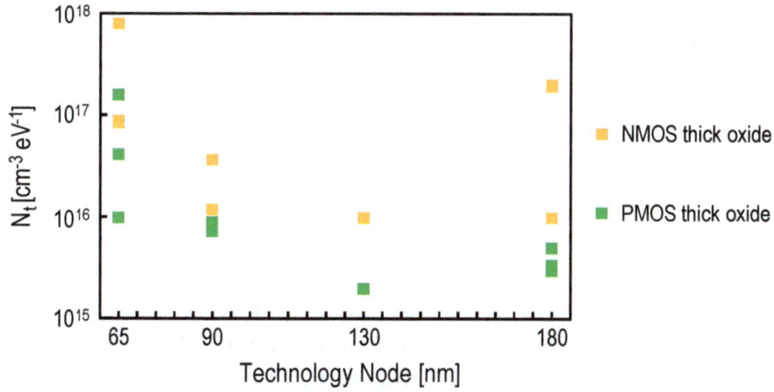

Fig. 5.6 The oxide trap density N_t, of PMOS and NMOS thick oxide transistors, as a function of the technology node based on data reported in design kits from different foundries

carriers in the channel are kept away from the silicon oxide interface [3]. It has been shown that using buried channel NMOS source followers leads to sub-electron noise performance [4].

From a design perspective, thick-oxide transistors, that operate at voltages as high as 3.3 V, are commonly used in CIS pixels. Figure 5.6 shows the oxide trap density of PMOS and NMOS thick oxide transistors from different foundries and technology nodes. It shows that PMOS transistors feature, generally, an oxide trap density lower than NMOS transistors. Using an in-pixel PMOS source follower transistor also led to a sub-electron noise performance [5]. The drawback of using an in-pixel PMOS transistor is the reduction of the fill factor due to the spacings imposed by the layout design rules and the possible quantum efficiency reduction if the PMOS n-well is too close to the PPD.

5.2.2.2 Reduce the Sense Node Junction and Parasitic Capacitance

The input referred $1/f$ noise expression (5.4) shows that the latter is linearly dependent on the capacitance C_P expressed in (4.10). C_P is the total capacitance resulting from the floating diffusion junction and the parasitic capacitances connected to the sense node including the overlaps with the transfer and reset gates as well as the metal wires coupling capacitances. The reduction of C_P results directly in a increased conversion gain (4.13) which also reduces the thermal noise (5.1) at the cost of a lower pixel dynamic range. The noise reduction through C_P minimization is an active research topic. Careful layout is not enough to significantly decrease the contributions of the transfer and reset overlap capacitances as well as the junction capacitance of the floating diffusion. To this purpose, process improvements are necessary. Indeed, recent works presenting sub-electron readout noise CIS focused on this point. In [6], different process level techniques have been presented leading to the reduction of C_P. It has been shown that the omission of the low doped drains (LDDs) used in standard CMOS transistors reduces effectively the gate overlap capacitances. Also, increasing the depletion depth under the floating diffusion by reducing the doping concentration reduces the junction capacitance. The combination of these techniques led to a C_P reduction of about 47%. In [7], the capacitance C_P has been reduced by using an idea called "virtual phase" well known in CCDs consisting in creating a potential profile that isolates the floating diffusion from the transfer gate. In this way, the overlap capacitance between the transfer gate and the floating diffusion (denoted C_{Tov} in Fig. 4.8) is dramatically reduced. Furthermore, the channel width of the reset transistor is reduced by controlling the doping profile in order to reduce the overlap between the reset gate and the floating diffusion (denoted C_{Rov} in Fig. 4.8). But this was obtained at the cost of a low pixel full well capacity and a relatively higher lag. In [8], C_{Tov} is reduced by introducing a special implant isolating the transfer gate from the SN and C_{Rov} is reduced by omitting the reset transistor. But this requires the reset to be performed using an off-chip high voltage clock of 25 V connected directly to an implant close to the SN. In this way, the sense node is reset by a punch through effect. Figure 6.2 shows the impact of the C_P reduction through a plot of the calculated input-referred $1/f$ noise as a function of the gate width and length, based on (5.4), for a C_P of 0.75 fF corresponding to a standard process and a C_P of 0.25 fF corresponding to the one that could be obtained through advanced process refinements [6]. Figure 5.7 shows the effectiveness of this C_P reduction which leads to a reduction of the input-referred $1/f$ noise from 0.4 to below 0.3 e^-_{rms}.

5.2.2.3 Use a Minimum Gate Width and an Optimal Length

Based on (5.4), it can be shown analytically [9] or numerically using the a plot of (5.4) versus the gate width and length, that the lowest input-referred $1/f$ noise corresponds to the minimum gate width and an optimal length generally slightly higher [9]. Indeed, the expression of the input-referred $1/f$ noise is proportional to the function $F(W, L)$ defined as

$$F(W, L) = \frac{1}{W \cdot L}(C_P + \frac{2}{3}C_{ox} \cdot W \cdot L + C_e W)^2 \tag{5.5}$$

The optimal W and L correspond theoretically to the minimum of $F(W, L)$. For this two variables function, the local and global minimums correspond to the points where the gradient of $F(W, L)$ is nul. Otherwise the minimum exists in the borders of the era $[W_{min}, +\infty[\times[L_{min}, +\infty[$, where W_{min} and L_{min} are the minimum width and length allowed by the technology. It can be shown that

$$\vec{\nabla} F(W, L) = 0 \Leftrightarrow L = -\frac{1}{\frac{2}{3}C_{ox}}\left(\frac{C_P}{W} + C_e\right) \tag{5.6}$$

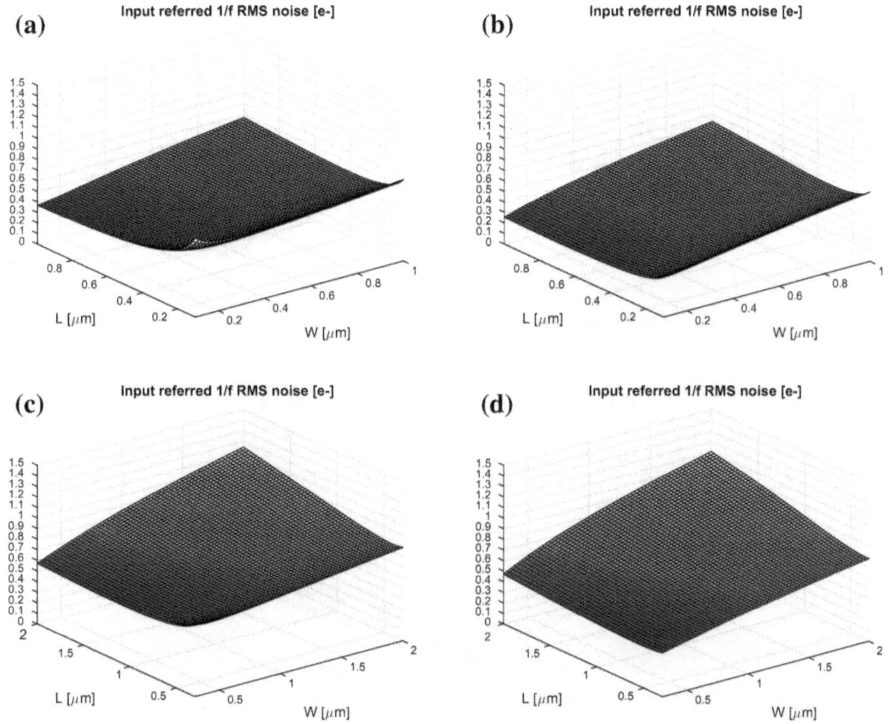

Fig. 5.7 The calculated input-referred $1/f$ noise, based on Eq. 5.4, as a function of the in-pixel source follower width W and length L with $\alpha_{CMS} = 3.5$ and $K_F = 10^{-27}\,\mathrm{F^2V^2m^{-2}}$. **a** corresponds to a thin oxide source follower with $C_e = 0.95\,\mathrm{fF/\mu m}$, $C_{ox} = 9.5\,\mathrm{fF/\mu m^2}$ and $C_P = 0.75\,\mathrm{fF}$. **b** corresponds to the same thin oxide source follower transistor and $C_P = 0.25\,\mathrm{fF}$ to illustrate the impact of the C_P reduction. **c** corresponds to a thick oxide source follower with $C_e = 0.45\,\mathrm{fF/\mu m}$, $C_{ox} = 4.5\,\mathrm{fF/\mu m^2}$ and $C_P = 0.75\,\mathrm{fF}$. **d** corresponds to the same thick oxide source follower transistor and $C_P = 0.25\,\mathrm{fF}$ to illustrate the impact of the C_P reduction when using a thick oxide transistor. Comparing (**a**) and (**b**) with (**c**) and (**d**) shows the interest of using thin oxide transistors

L is positive, consequently, the gradient of $F(W, L)$ cannot be nul. Hence, the minimum of $F(W, L)$ exists in the border corresponding to $W = W_{\min}$ or $L = L_{\min}$. It can be shown that

$$\frac{\partial}{\partial L} F(W_{\min}, L) = 0 \Leftrightarrow L = \frac{C_e W_{\min} + C_P}{\frac{2}{3} C_{ox} W_{\min}} \tag{5.7}$$

and in the same way

$$\frac{\partial}{\partial L} F(W, L_{\min}) = 0 \Leftrightarrow W = \frac{C_P}{\frac{2}{3} C_{ox} W_{\min} + C_e} \tag{5.8}$$

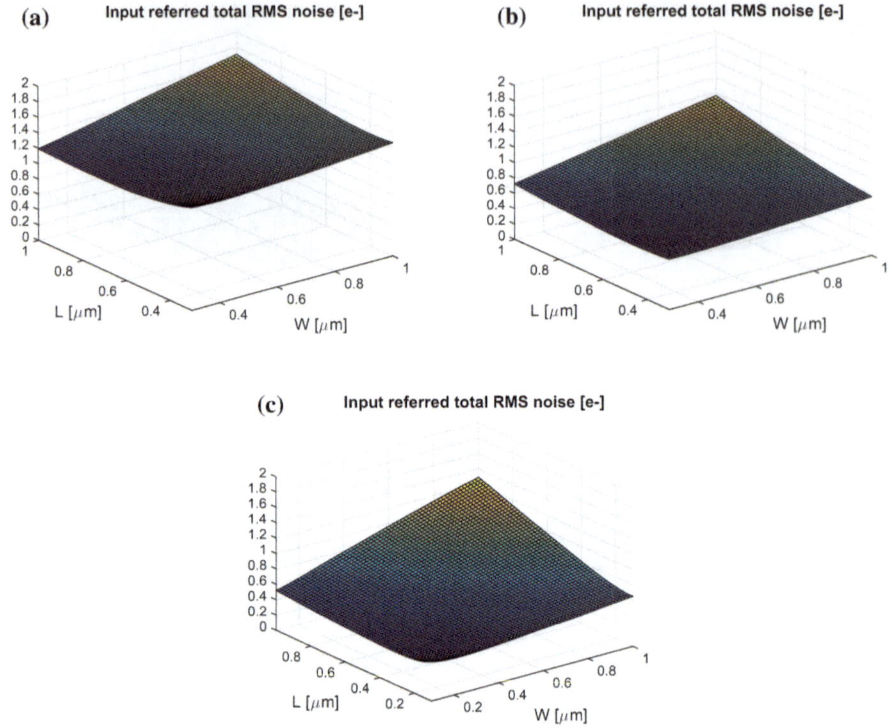

Fig. 5.8 The calculated input-referred total noise for a thick oxide NMOS, a thich oxide PMOS and a thin oxide PMOS source follower, based on (5.4) and (5.1), as a function of the in-pixel source follower width W and length L with $\alpha_{CMS} = 4.5$ corresponding to a simple CDS. **a** corresponds to a thick oxide NMOS source follower with $C_e = 0.45\,\text{fF}/\mu\text{m}$, $C_{ox} = 4.5\,\text{fF}/\mu\text{m}^2$, $C_P = 0.75\,\text{fF}$ and $N_t = 1.5\,10^{17}\,\text{eV}^-1\text{cm}^{-3}$ corresponding to $K_F = 10^{-26}\,\text{F}^2\text{V}^2\text{m}^{-2}$. **b** corresponds to a thick oxide PMOS source follower with $C_e = 0.45\,\text{fF}/\mu\text{m}$, $C_{ox} = 4.5\,\text{fF}/\mu\text{m}^2$, $C_P = 0.75\,\text{fF}$ and $N_t = 3\,10^{16}\,\text{eV}^-1\text{cm}^{-3}$ corresponding to $K_F = 2\,10^{-27}\,\text{F}^2\text{V}^2\text{m}^{-2}$. **c** corresponds to a thin oxide PMOS source follower from the digital library with $C_e = 0.95\,\text{fF}/\mu\text{m}$, $C_{ox} = 9.55\,\text{fF}/\mu\text{m}^2$, $C_P = 0.75\,\text{fF}$ and $N_t = 1.5\,10^{16}\,\text{eV}^{-1}\text{cm}^{-3}$ corresponding to $K_F = 10^{-27}\,\text{F}^2\text{V}^2\text{m}^{-2}$. The minimum width and length for thick oxide transistors is 0.3 and 0.1 μm for thin oxide transistors

By calculating the difference between the values of $F(W, L)$ in the minima points defined in (5.7) and (5.8). One can find that this difference has the same sign as $C_P - \frac{2}{3}C_{ox}W_{min}L_{min}$. As shown in the previous section, typical values of C_P in standard processes is about 0.75 fF. It can be lowered to values close to 0.25 fF with process refinements. The term $\frac{2}{3}C_{ox}W_{min}L_{min}$ is for instance in a 180 nm process of the order of 0.14 fF for W_{min} and L_{min} of 0.2 μm and C_{ox} of 4.5 $\text{fF}/\mu\text{m}^2$. Hence, the expression $C_P - \frac{2}{3}C_{ox}W_{min}L_{min}$ is generally positive. Consequently, the minimum input-referred $1/f$ noise corresponds to the minimum width allowed by the technology and the optimal length given by (5.7).

Figure 5.8 illustrates this principle on practical examples. It shows a plot of the input-referred $1/f$ noise as a function of the gate width and length for different combinations of C_P and source follower capacitance densities in a 180 nm CIS process. Figure 5.7a, b show the case of a source follower with an thin oxide source follower corresponding to a C_{ox} of 9.5 fF/μm^2 and minimum width and lengths of 0.2 μm for C_P of 0.75 and 0.25 fF. Figure 5.7c, d show the case of a source follower with an thick oxide source follower corresponding to a C_{ox} of 4.5 fF/μm^2 and minimum width and length of 0.3 μm for the same values of C_P. All these practical examples confirm that the minimum $1/f$ noise corresponds to the minimum gate width. The optimal gate length might be much higher is some cases. But the variation of the input-referred $1/f$ noise over L becomes very week starting from a slightly higher L.

5.2.2.4 Increase the Oxide Capacitance per Unit Area of the Readout Transistor

Based on (5.4), the remaining parameter involved in the input referred $1/f$ noise expression is the in-pixel amplifying transistor oxide capacitance per unit area C_{ox}. In case the the gate oxide capacitance of the in-pixel readout transistor dominates the other capacitances connected to the sense node, C_{ox} has no significant impact on the input referred $1/f$ noise. But in general, the capacitance C_P including the floating diffusion junction and other parasitic capacitances is higher than the gate oxide capacitance given by $\frac{2}{3}C_{ox}WL$. Consequently, a higher C_{ox} is expected to result in a lower input-referred $1/f$ noise even with a lower conversion gain.

Standard libraries generally offer a selection of transistors featuring different oxide thicknesses and consequently different oxide capacitance densities. Generally, thin oxide transistors support low voltages and they derive from the digital library. Thick oxide transistors support higher voltages. They can be used for analog blocks and input-output signal conditioning. Thus, in most CIS processes, all the gates included in the pixel feature thick oxides since the transfer gate and the reset gate are controlled by high voltages (3.3 V), the SF is also chosen as a thick oxide transistor to exploit a high dynamic range.

Hence, the above analysis shows that if the conventional choice of a thick oxide in-pixel amplifying transistor is intuitive given the high voltages used in the pixel, it is not the optimal choice for the $1/f$ noise performance. The parameter C_{ox} is a degree of freedom that has not been exploited for $1/f$ noise reduction in CIS. The following section suggests a practical way to use this parameter in a standard process and shows that other advantages results from this choice. The next chapter demonstrates this idea with measurement results.

5.3 Thin Oxide Source Follower: A Good Match

From the designer's perspective, the implementation of a thin oxide source follower offer both a higher oxide capacitance per unit area and a lower gate width. A thin oxide SF with a minimum gate width features also lower overlap capacitances. It is important to verify that the choice of a thin oxide transistor does not come at the cost of a negative impact on the $1/f$ noise process parameter K_F. A thinner oxide is expected to feature a better control of the gate over the channel and therefor a lower K_F. The oxide trap density of thin oxide PMOS transistors and thick oxide NMOS transistors of different foundries and technology nodes have been compared in [10] based on data reported in design kits. This comparison showed that the thin oxide PMOS transistors generally feature a lower oxide trap density. Thus, using a thin oxide PMOS source follower is suited for all the pixel-level noise reduction mechanisms described above and related to the source follower.

Consider the example of the 180 nm CIS process used in this work. The design kit offers transistors with two oxide thicknesses. The thick oxide NMOS transistor has an oxide capacitance per unit area $C_{ox} = 4.5\,\text{fF}/\mu\text{m}^2$ with a minimum gate width close to 0.4 μm and an oxide trap density obtained from the BSIM noise parameter NOIA of $N_t = 1.5\,10^{17}\,\text{eV}^{-1}\text{cm}^{-3}$, while the thick oxide PMOS feature the same parameters with a lower trap density of $N_t = 2\,10^{16}\,\text{eV}^{-1}\text{cm}^{-3}$. The thin oxide PMOS transistor has a higher oxide capacitance density of $C_{ox} = 9.55\,\text{fF}/\mu\text{m}^2$, a lower minimum width close to 0.2 μm and a lower oxide trap density of $N_t = 1.5\,10^{16}\,\text{eV}^{-1}\text{cm}^{-3}$. Hence, theoretically, the thin oxide PMOS source follower is the optimal choice. In order to illustrate the difference between the three source followers in terms of the noise performance, Fig. 5.8 shows plots of the input-referred total noise calculated using (5.4) and (5.1) using the parameters of each transistor type. It shows the clear advantage of using thin oxide PMOS source followers that comes with an RMS noise about three times lower than a traditional thick oxide NMOS.

In order to validate this idea in the simulator, a transient noise simulation is performed on three readout chains based respectively on a standard thick oxide NMOS, a thick oxide PMOS and a thin oxide PMOS source follower in the 180 nm CIS process mentioned above. The compared readout chains share the same column level amplification and a simple CDS. Figure 5.9 shows the impact of the in-pixel source follower transistor type on the input-referred $1/f$ noise. The PMOS SF based readout chain features a lower $1/f$ noise than the NMOS based one thanks to the increase of C_{ox} and the reduction of the parameter K_F. The readout chain based on the thin oxide PMOS source follower features the lowest input-referred noise because it cumulates a lower K_F, a higher oxide thickness and a lower minimum width. The simulation results of Fig. 5.9 match well the calculated input-referred noise and confirm the analytical results.

It is important to verify that these techniques are not harmful in terms of the thermal noise. Equation 4.24 shows that the thermal noise is reduced by increasing the conversion gain. Thus, using a thin oxide SF with smaller gate dimensions is also expected to reduce the input-referred thermal noise.

In order to predict the impact of the combination of thin oxide SF with process optimizations reducing the sense node capacitance, the parameter C_P of (5.4) is replaced by the measurement results from [6]. The starting point corresponds to the result of $0.4\,e^-_{rms}$ obtained in the simulation and corresponding to the case of a thin oxide source-follower based pixel design with standard rules. Then the values of C_P based on [6] are used to predict the evolution of the input-referred noise for

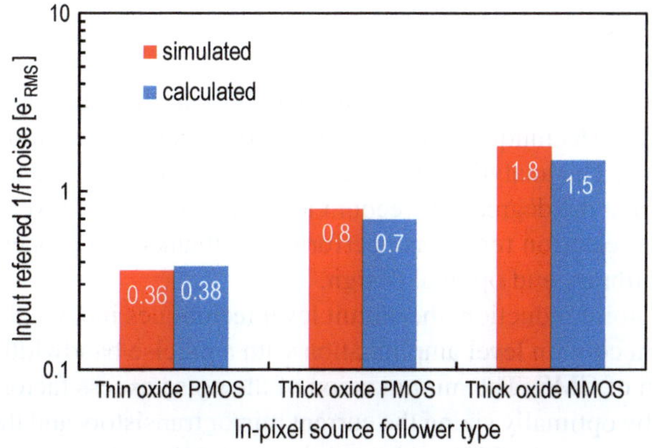

Fig. 5.9 Simulated input referred $1/f$ noise for three readout chains sharing the same column-level circuitry and using different in-pixel SF transistor types

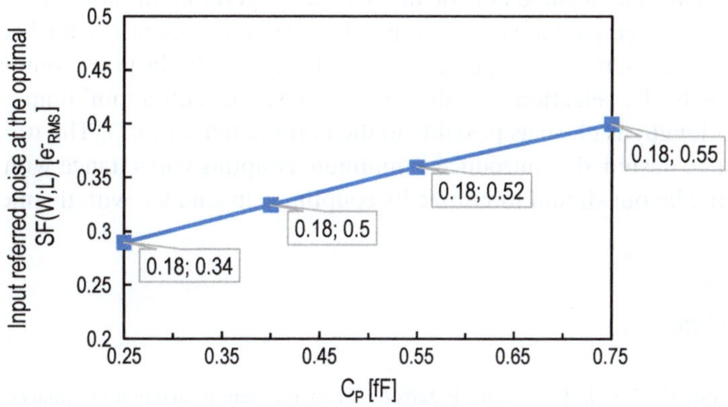

Fig. 5.10 The input-referred 1/f noise as a function of the capacitance C_P determined by the floating diffusion, wiring and parasitic capacitances independent from the source follower transistor. $C_P = 0.75$ fF corresponds to the case of a conventional sense node junction, transfer gate and transistors. $C_P = 0.55$ fF corresponds to the case of transfer and reset gates without low doped drains [6]. $C_P = 0.4$ fF corresponds to the case of transfer and reset gates without low doped drains and a sense node without channel stop underneath [6]. $C_P = 0.25$ fF corresponds to the case of more advanced process refinements as [6, 7]. The numbers in labels correspond to the minimum width and optimum length as discussed in 5.2.2.3

each additional technique used to reduce the SN capacitance. The result is plotted in Fig. 5.10. It shows the expected input-referred noise, at the optimal gate width and length, for each value of C_P. Figure 5.10 shows that the 0.3 e^-_{rms} limit can be crossed if the thin oxide PMOS SF is combined with process optimizations reducing C_P in the 180 nm process used in [10].

5.4 Summary

The input-referred read noise in conventional CIS can be reduced dramatically by combining multiple techniques. These include circuit design techniques and process-level refinements. The analytical calculations, confirmed by the transient noise simulations, show that the degrees of freedom left to the designer, if well exploited, can lead to deep sub-electron read noise performance thanks to a proper device choice in the standard library and optimal design.

For thermal noise reduction, the circuit level techniques include the implementation of high gain column level amplification with a precise bandwidth control or the implementation of CMS. The minimization of the noise excess factors of the source follower stage by optimally sizing the current mirror transistors and the column level amplifier by designing the OTA with a minimum number of noisy transistors.

By implementing the above techniques, the $1/f$ noise becomes the dominant read noise source. The reduction of the $1/f$ noise starts by the implementation of CDS. When the transistors located outside the pixel and operating in saturation regime are designed with large gate areas, the in-pixel source follower transistor becomes the dominant noise source. The optimal device choice for the in-pixel source follower corresponds to the selection of a thin oxide transistor with a minimum gate width and higher length as close as possible to the optimal length (5.7). The transfer reset gates must be designed to introduce a minimum coupling capacitance with the sense node and the layout should minimize its coupling capacitance with the metal wires.

References

1. P. Bolcato, M. Tawfik, R. Poujois, P. Jarron, A new efficient transient noise analysis technique for simulation of ccd image sensors or particle detectors, in *Proceedings of the IEEE Custom Integrated Circuits Conference 1993* (1993), pp. 14.8.1–14.8.4
2. A. Boukhayma, A. Peizerat, C. Enz, A correlated multiple sampling passive switched capacitor circuit for low light cmos image sensors, in *2015 International Conference on Noise and Fluctuations (ICNF)* (2015), pp. 1–4
3. Y. Chen, X. Wang, A. Mierop, A. Theuwissen, A cmos image sensor with in-pixel buried-channel source follower and optimized row selector. IEEE Trans. Electron Devices **56**(11), 2390–2397 (2009)
4. Y. Chen, Y. Xu, Y. Chae, A. Mierop, X. Wang, A. Theuwissen, A 0.7e-rms temporal-readout-noise cmos image sensor for low-light-level imaging, in *2012 IEEE International Solid-State Circuits Conference Digest of Technical Papers (ISSCC)* (2012), pp. 384–386
5. C. Lotto, P. Seitz, T. Baechler, A sub-electron readout noise cmos image sensor with pixel-level open-loop voltage amplification, in *2011 IEEE International Solid-State Circuits Conference* (2011) pp. 402–404
6. K. Fumiaki, W. Shunichi, N. Satoshi, K. Rihito, S. Shigetoshi, Analysis and reduction of floating diffusion capacitance components of cmos image sensor for photon-countable sensitivity, in *International Image Sensors Workshop* (2015)
7. J. Ma, D. Starkey, A. Rao, K. Odame, E. Fossum, Characterization of quanta image sensor pump-gate jots with deep sub-electron read noise. IEEE J. Electron Devices Soc. **3**(6), 472–480 (2015)

8. M. Seo, S. Kawahito, K. Kagawa, K. Yasutomi, A $0.27e^{-}_{rms}$ read noise $220\mu v/e^{-}$ conversion gain reset-gate-less cmos image sensor with $0.11 - \mu m$ cis process. IEEE Electron Device Lett. **36**(12), 1344–1347 (2015)
9. A. Boukhayma, A. Peizerat, A. Dupret, C. Enz, Design optimization for low light cmos image sensors readout chain, in *2014 IEEE 12th International New Circuits and Systems Conference (NEWCAS)* (2014), pp. 241–244
10. A. Boukhayma, A. Peizerat, C. Enz, Temporal readout noise analysis and reduction techniques for low-light cmos image sensors. IEEE Trans. Electron Devices **63**(1), 72–78 (2016)

Chapter 6
Design of a Sub-electron Readout Noise Pixel in a Standard CIS Process

Different noise reduction approaches, at process and design level, have been suggested in the previous chapter. The process level modifications can be performed in order to reduce the sense node capacitance as well as the overlap capacitances of the transfer and reset gates. Such modifications require a good knowledge of the process parameters and a flexible foundry willing to modify the standard process. The purpose of this chapter is to exploit all the degrees of freedom left to the designer (and their are not that many!) in order to reach the optimal noise performance in a standard process without any process-level refinements. It has been shown in the previous chapter that the choice of an in-pixel thin oxide source follower instead of the traditionally used thick oxide transistors is expected to reduce significantly the input-referred noise. In this chapter, we present the first test chip, to our knowledge, embedding pixels with an optimally sized thin oxide source follower and integrated in a standard 180 nm CIS process. In order to validate the efficiency of this noise reduction technique, the newly proposed pixel is compared to a state of the art pixel proposed by the foundry embedding a buried channel thick oxide SF.

6.1 The Optimized Pixel Design

As discussed in Chap. 5, the $1/f$ noise originating from the SF transistor can be reduced using a SF with a thinner oxide and lower gate width. From a designer perspective, this can be achieved by using a thin oxide (1.8 V) transistor instead of a thick oxide (3.3 V) in the 180 nm CMOS process used for this work. The sense node needs to be reset at a voltage higher than 2 V for a good dynamic range and also to make sure that the sense node attracts efficiently the charges from the PPD when the potential barrier under the transfer gate is removed. Yet, the voltage difference between the gate and the bulk of the thin oxide SF must remain smaller than 1.8 V. Hence, the bulk voltage of the thin oxide SF must be shifted in order to keep the gate-to-bulk voltage below 1.8 V. To this purpose a PMOS thin oxide transistor is chosen. Indeed, the bulk voltage of a PMOS transistor can be controlled through the

A. Boukhayma, *Ultra Low Noise CMOS Image Sensors*,
Springer Theses, https://doi.org/10.1007/978-3-319-68774-2_6

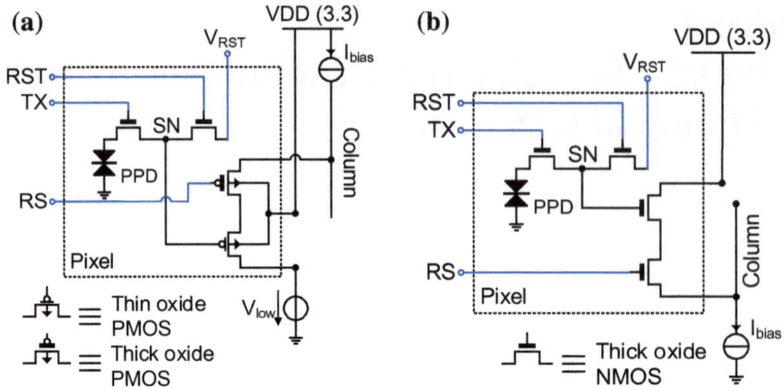

Fig. 6.1 Schematic of the proposed pixel (**a**) compared to a conventional pixel (**b**)

n-well connection. An NMOS transistor with a separated p-well could also be used. Such an option in not available in all foundries especially the ones dedicated for image sensors. additionally, a PMOS thin oxide SF usually shows a lower $1/f$ noise parameter K_F than NMOS transistors in the targeted technology node. Consequently, the K_F parameter in (4.28) is expected to be lower for the PMOS thin oxide SF. A thick oxide PMOS row select (RS) transistor is chosen in order to put it in the same n-well as the SF and optimize the layout footprint of the in-pixel readout transistors. Figure 6.1a shows the schematic of the proposed pixel. The bulk voltage of the thin oxide PMOS SF is set to VDD at 3.3 V to insure that the voltage difference between its bulk and the gate (sense node) is positive and below 1.8 V. The drain voltage of the SF is connected to (V_{low}) which is set to 1.5 V in order to shift the ground of the SF stage and make sure that the voltage differences between all the thin oxide SF terminals remain below 1.8 V. The SF gate is sized to minimize its input-referred temporal read noise (TRN) according to (4.28), which is plotted in Fig. 6.2 versus W and L. It clearly shows that W should be chosen as mimum ($W = 0.22\,\mu$m) whereas L should be chosen slightly higher ($L = 0.24\,\mu$m) in order to minimize the input-referred $1/f$ TRN of the SF to about 0.4 e^-_{rms}. The transfer (TX) gate and reset (RST) transistors are thick oxide NMOS.

6.2 Test Chip Architecture

In order to confirm the theoretical results presented in the previous sections, a test chip has been designed to compare a state-of-the-art pixel, already optimized for low noise, with the newly proposed pixel based on the thin oxide PMOS source follower introduced in Fig. 6.1a. Both pixels are based on 4 transistors and a classical pinned photo diode. Figure 6.3 shows the layouts of the proposed pixel and the reference pixel. The new pixel features a size of 7.5 μm × 6.5 μm and a fill factor of 66%.

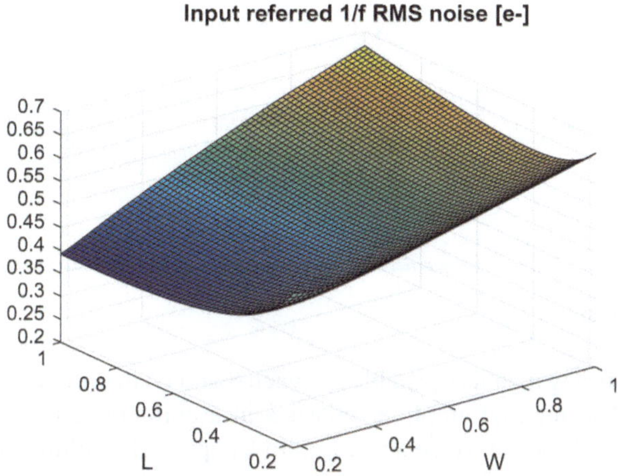

Fig. 6.2 Calculated input-referred $1/f$ noise using (4.28) as a function of the thin oxide SF gate size

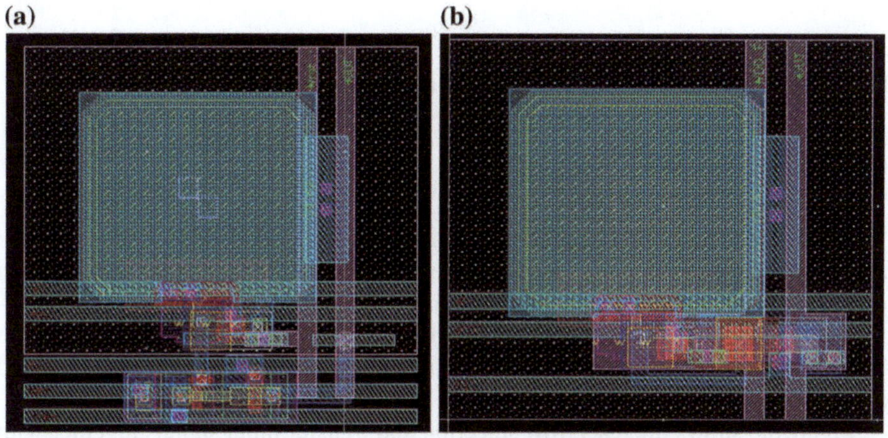

Fig. 6.3 Layout of the proposed pixel (**a**) compared to the reference pixel (**b**)

The layout of the newly proposed pixel shows a spacing between the PPD with the transfer and reset gate and the SF and RS PMOS transistors put in the same n-well. This spacing is imposed by the design rules of the foundry. Hence, the reference pixel used for comparison features a slightly smaller area of 6.5 μm $\times$ 6.5 μm for the same active area.

A 5 mm $\times$ 1 mm chip has been designed in order to test readout chains based on the two presented pixels. Figure 6.4 presents the overall architecture of the test chips. Each chip includes a total of 24 columns connected to small arrays of the new and reference pixels, each surrounded by 8 dummy pixels for proper characterization. Both pixels are connected to the same column-level amplifiers limiting the bandwidth to

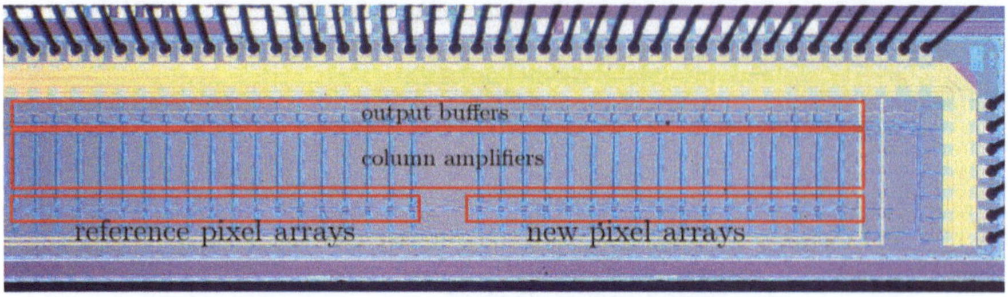

Fig. 6.4 Test chip micrograph

265 kHz and offering a gain adjustable between 8 and 64 in order to check its impact on the noise. The internal amplifier consists in a common source fully cascoded amplifier designed for low noise according to Chap. 5. The column amplifier is followed by a voltage buffer to drive the signal to the input of an external 14 bits ADC. Figure 6.4 shows a micrograph of the chip designed in a standard 180 nm CIS process.

6.3 Test Description and Measurement Results

6.3.1 Test Setup

A printed circuit board (PCB) has been designed (Fig. 6.5) and fabricated in order to drive each readout column comprising the reference and new pixels. Each column output is connected to its own pad. The multiplexing between the columns is performed on-board. An FPGA is used in order to generate the signals controlling the pixels and the column-level amplifier, namely, the transfer, reset, row selection and auto-zero phases.

Figure 6.6 shows the reset, row selection and transfer input signals together with the output of single column based on the newly proposed pixel measured with a scope. The pixel is selected when the row selection signal (red) is down since the RS switch is a PMOS. The sense node is immediately reset by setting the reset signal to 3.3 V. The charge transfer from the PPD to the sense node is performed by applying a voltage of 2.7 V on the transfer gate (green) during 0.5 μs. The value of 2.7 V is chosen in order to make the voltage under the transfer gate lower than the sense node voltage (about 2.7 V) and higher than the pin voltage (estimated to be about 1 V). In this way, the electrons do not remain under the transfer gate during the charge transfer. Once the transfer is activated, the output signal level (blue) rises as the electrons transfer to the sense node. The signal establishment takes about 5 μs, which is compilent with the bandwidth of 265 kHz set by the column-level amplifier. Note that the column-level amplifier designed in this work is an inverter. Thus, the voltage at the level of the sense node decreases while the output of the column-level amplifier increases after the charge transfer from the sense node to the PPD.

Fig. 6.5 Image of the tested chip with the test PCB

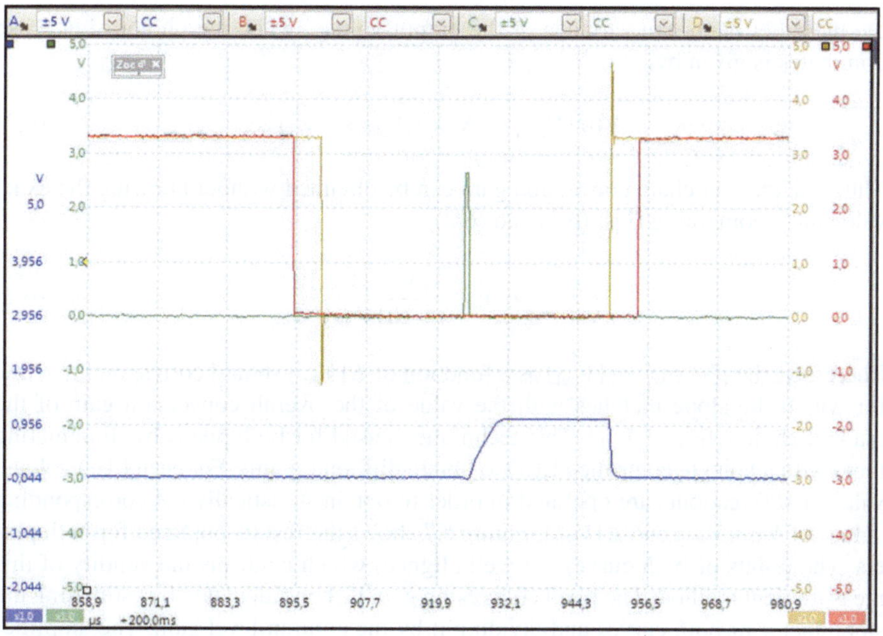

Fig. 6.6 A screen shot from the scope window showing the row selection signal (red), the reset signal (brown), the transfer signal (green) as well as the chip output of a single pixel (blue) based on the thin-oxide PMOS SF

The test follows the same steps used for the noise calculation. The overall conversion gain ($A_{CG} \times A_{col}$) of each readout chain is first measured, then the output noise is measured and referred to the input using the overall conversion gain. The source follower and column amplification voltage gain is measured using the reset voltage as an input while closing the reset switch.

6.3.2 Conversion Gain Measurement

In order to measure the conversion gain, the photon transfer curve (PTC) method [1] is used. The photon shot noise dominates the readout noise when the pixel receives an amount of photons between the noise floor and saturation. For an average number N of received photons, the variance of the corresponding shot noise is given by N. On the one hand, the mean value of the signal at the output of the readout chain is given by

$$E[V_{out}] = N \times A_{CG} \times A_{col}. \tag{6.1}$$

On the other hand, the variance of the output voltage when the photon shot noise dominates is given by

$$Var[V_{out}] = N \times (A_{CG} \times A_{col})^2. \tag{6.2}$$

Thus, the readout chain conversion gain can be obtained without knowing the exact value of N combining Eqs. (6.1) and (6.2)

$$A_{CG} \times A_{col} = \frac{Var[V_{out}]}{E[V_{out}]}. \tag{6.3}$$

Therefore, the plot of $Var[V_{out}]$ as a function of $E[V_{out}]$ should correspond to a line for which the slope matches with the value of the overall conversion gain of the readout chain ($A_{CG} \times A_{col}$). This technique is used for both pixels by illuminating them with a voltage controlled LED to obtain different points. For each LED voltage value, a 100 readouts are operated in order to obtain statistically the corresponding values of $Var[V_{out}]$ and $E[V_{out}]$. Figure 6.7 shows the results obtained for both pixels. The points of both curves are well aligned, which confirms the validity of this measurement method. The pixel conversion gain is then calculated by estimating the slope factor of both curves and dividing it by the column-level gain. The amplifier was designed to provide two gain configuration $\times 8$ and $\times 64$. The measured overall gain including the output buffer of the PCB for the $\times 8$ configuration is 7. Thus the conversion gain of the new pixel is approximately equal to 185 $\mu V/e^-$ whereas for the reference pixel, it is close to 85 $\mu V/e^-$.

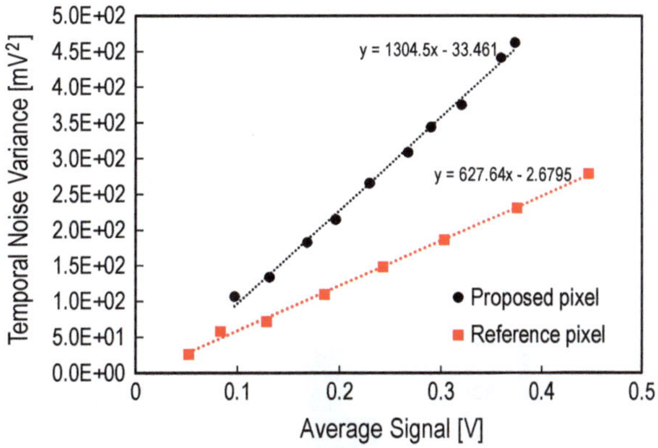

Fig. 6.7 Measurement of the conversion gain using the variance-mean photon transfer curve

6.3.3 Input-Referred Read Noise

The output noise variance and RMS value are measured for all pixels by using the 14 bits ADC of the scope. The noise is measured by operating 1000 readouts for each pixel without activating the transfer gate. For the input-referred noise measurement, the output noise in mV_{rms} is first measured, then referred to the input using the overall conversion gain of the readout chain including the pixel, the column-level amplifier and the output buffer of the PCB. The average overall conversion gain measured with the PTC curve for the high gain configuration is 9.1 mV/e^- for the new pixels and 4.4 mV/e^- for the conventional pixels. Figure 6.8 shows the histograms of the input-referred noise charge obtained for all the new and reference pixels (85 of each) with the high column gain configuration (×64). The new pixel shows a lower average input-referred noise below 0.4 e^-_{rms}. The reference NMOS source follower based pixel features an average input-referred noise of 0.9 e^-_{rms}. In addition, the histograms show that the standard deviation corresponding to the noise measurement of the new pixels is three times smaller than that of the reference pixel. The impact of the column-level gain on the average value of the input-referred noise for both pixels is shown in Fig. 6.9. The new pixel shows more than 50% less noise than the reference pixel even at low gain ($A_{col} = 8$) when $1/f$ noise is not dominating. Indeed, the thin oxide source follower features a higher conversion gain and features a gate width and length twice smaller than the thick oxide source follower, thus, based on (4.24), this result confirms the theoretical expectation mentioned in the previous chapter.

The comparison of the presented readout chain based on the new pixel with the state-of-the-art pixel shows that the proposed noise reduction technique is promising. Indeed, these results were obtained using a "digital" thin oxide PMOS which leaves some room for more optimization at the process level.

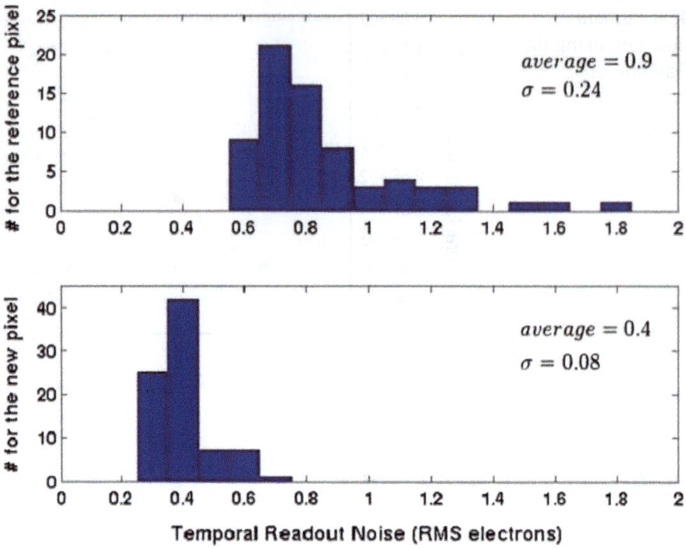

Fig. 6.8 Input-referred noise histograms measured for the newly proposed pixels and the reference pixels

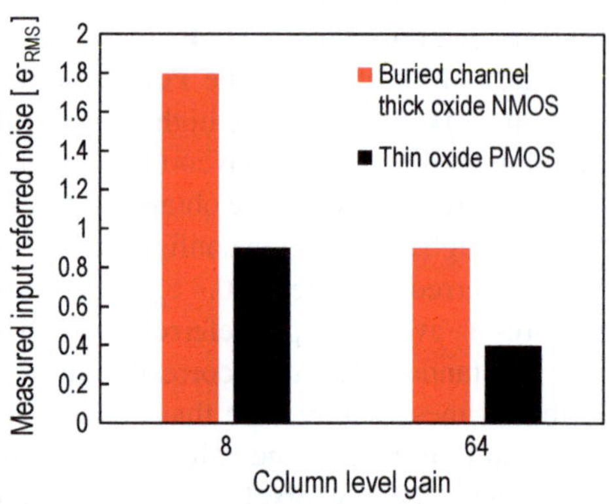

Fig. 6.9 Measured input-referred noise for the two column-level gain configurations

6.3.4 Pixel Voltage Swing

One of the main reasons of using in-pixel thick oxide SFs is the voltage swing that they offer. A voltage swing of the order of 1 V is enough for imaging applications. Indeed, with a conversion gain of about $100\,\mu V/e^{-}$, 1 V swing corresponds to 10,000 electrons which is high enough even for applications requiring relatively high dynamic range. In order to verify that the implementation of the thin-oxide source follower does not dramatically reduce the voltage swing at the output of the pixel, we propose the following measurement that also gives an estimation of the pinning voltage of the PPD. The exposure time of the pixel and the lighting level are adjusted

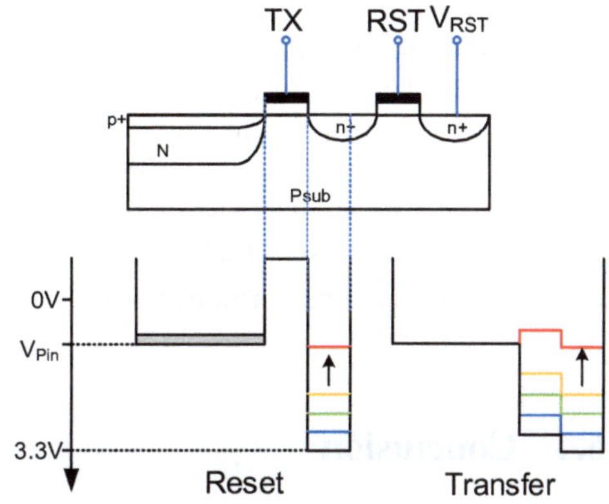

Fig. 6.10 Schematic of the experiment performed to estimate the pin voltage

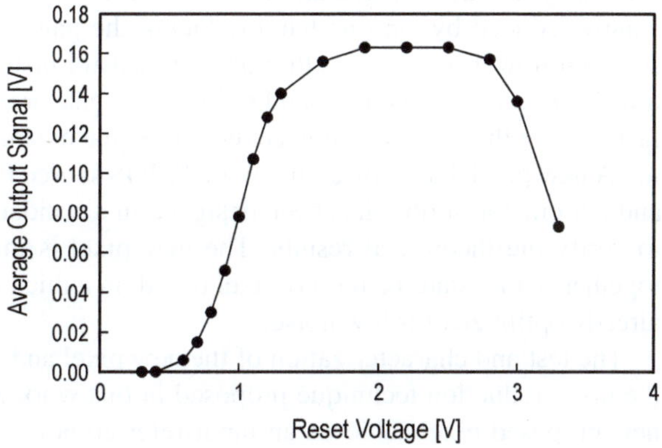

Fig. 6.11 Average output signal as a function of the reset voltage, of the newly proposed pixel, for a fixed exposure time and lighting level

in order to integrate a number of photoelectrons corresponding to an output signal level of a few tens of mV. The integration and readouts are performed several times and the outputs are averaged. This operation is then repeated by sweeping the reset voltage from 0 to 3.3 V. Figure 6.10 illustrates this operation. Theoretically, for all the reset voltage values higher than the pinning voltage of the PPD, the same amount of electrons is transferred to the sense node resulting in the same signal voltage at the output of the readout chain. When the reset voltage becomes closer to the pinning voltage level, the output signal starts decreasing until it completely decays when the reset voltage becomes lower than the pinning voltage in which case the electrons are not transferred from the PPD to the sense node since they see no potential well. Figure 6.11 shows the obtained curve when applying this measurement technique to the newly proposed pixel based on the thin oxide SF. It shows that the charge starts transferring to the sense node for a reset voltage of 0.5 V but an efficient charge transfer from the PPD occurs for reset voltages between 1.8 and 2.8 V. Hence, the source follower stage based on the thin oxide PMOS transistor allows a voltage swing of at least 1 V. The in-pixel thin oxide source follower behaves as expected theoretically. Indeed, the voltage at its node is not supposed to be lower than 1.5 V in order to maintain a gate-to-bulk and a gate-to-source voltages below 1.8 V which is the nominal limit for this thin oxide. One can see that the output voltage of the pixel also decays when the reset voltage goes higher than 2.8 V. Indeed, when the sense node voltage is about 3 V the voltage at the source level becomes much close to 3.3 V. This is due to the fact that, for a PMOS SF, the source to gate voltage corresponds roughly to the threshold voltage which is around 300 mV in this technology. Since the bias is provided by a MOS based current mirror, the latter is no more saturated which degrades the performance of the SF.

6.4 Conclusion

The analytical noise calculation shows that the input-referred noise can be significantly reduced by, on one hand reducing the gate oxide thickness of the in-pixel source follower (choosing a thin oxide transistor instead of a traditional thick oxide transistor) and on the other hand reducing its gate area, assuming that the $1/f$ noise factor K_F of the thin oxide device remains equal or smaller than a thick oxide transistor. A new pixel, based on a thin oxide PMOS source follower with a pitch of 7.5 µm and a fill factor of 66% has been designed in a standard 180 nm CIS CMOS process to verify the theoretical results. The new pixel is implemented on the same chip together with a state-of-the-art pixel based on a thick oxide NMOS source follower already optimized for low noise.

The test and characterization of the new pixel and the reference pixel shows that the noise reduction technique proposed in this work is very promising. Indeed, the new proposed pixel achieves an input-referred noise of $0.4\,e^-_{rms}$ corresponding to a noise reduction of more than a factor 2 compared to the classical reference pixel which features a $0.9\,e^-_{rms}$ input-referred noise. But these results are obtained on a small number of pixels. In the next chapter, it will be shown how the pixel layout can be further improved and how well this proposed pixel design performs in a full VGA imager.

Reference

1. J.R. Janesick, *Photon Transfer* (SPIE, USA, 2007), pp. 49–79

Chapter 7
Characterization of a Sub-electron Readout Noise VGA Imager in a Standard CIS Process

In the previous Chapter, we demonstrated that sub-electron read noise performance can be achieved, in a standard CIS process, at room temperature by optimal design. The design optimization included, for the first time, the choice of a thin oxide PMOS SF. These measurements gave promising results but on a small number of isolated pixels. This chapter presents the first implementation, in a standard process, of pixels with thin oxide PMOS SFs in a full VGA (640H×480V) image sensor. The proposed imager features an input-referred noise histogram ranging from 0.25 e^-_{rms} to a few e^-_{rms} and peaking at 0.48 e^-_{rms}. In the mean time, it features an improved pixel layout leading to a pitch of 6.5 μm, a dynamic range of 82.5 dB corresponding to a full well capacity of 6400 e^- and a frame rate up to 80 fps. This imager crosses the bridge between highly sensitive low-light CIS and conventional imagers. It demonstrates the efficiency of the proposed design level noise reduction technique that can be easily combined with the process optimizations mentioned in Chap. 5 for even more noise reduction.

This chapter presents, in the first section, the overall architecture of the imager as well as its most important blocks. In the second section, a detailed characterization of the proposed imager with a description of the measurement methods are presented. The test results are discussed in the third section.

7.1 Chip Architecture and Circuit Design

7.1.1 Imager Architecture

Figure 7.1 shows the overall architecture of the imager. It is a rolling shutter CMOS image sensor with 640(H)×480(V) array of the newly proposed 4 T pixels with standard CIS PPDs. A row control mixed signal block generates for each line of pixels the reset (RST), transfer (TX) and selection (RS) signals (Fig. 7.6). The row control block also allows the control of the integration time. Each column of the

A. Boukhayma, *Ultra Low Noise CMOS Image Sensors*,
Springer Theses, https://doi.org/10.1007/978-3-319-68774-2_7

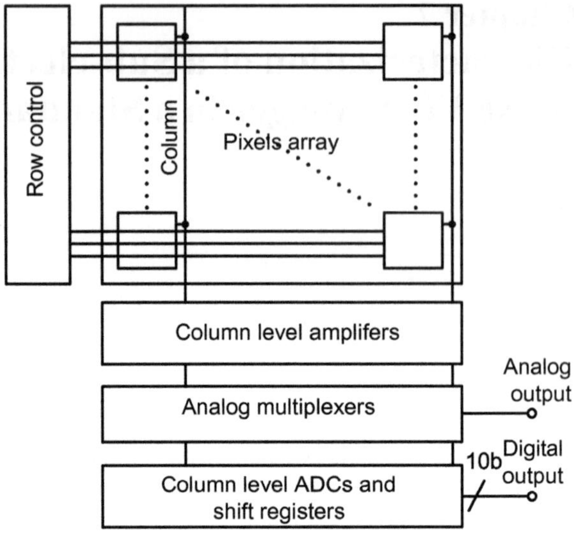

Fig. 7.1 Overall architecture of the proposed image sensor

pixels array is connected to a closed loop gain amplifier introducing gain and limiting the bandwidth. A mixed signal analog multiplexing block made of shift registers, voltage level shifters and analog switches is implemented at the output of the column-level amplifiers. It allows to choose between an analog or digital output. In the digital readout mode, each column-level output is simply connected to the input of a 10 bits single-slope ADC (SS-ADC). All the columns are read in parallel. In this configuration, the imager operates at 80 fps. In the analog readout mode, the shift registers and analog switches connect the columns to the analog output. In this case, the columns are read one after the other. The analog readout mode is important for proper characterization of the pixels. In this mode the frame rate is 640 times slower but the column and pixel readout time remain the same.

7.1.2 Pixel Design and Layout

The pixel design has been presented in the previous chapter. This section addresses the layout optimization. A minimum distance is imposed between the n-well and the neighboring NMOS and PPD. Also, a minimum distance between thin oxide and thick oxide transistors has to be fulfilled. In order to address these issues, an optimized layout still filling all the standard design rules is proposed. The optimization relies on putting the maximum number of transistors in the same n-well and keeping this latter away from the PPDs. Figure 7.2b shows the proposed layout of neighboring pixels. The repeated pattern of the pixels array consists of two symmetrical pixels sharing a common n-well with the two SFs and RSs. Inside the N well, a common thin oxide area contains the two SFs. This compact layout results in a fill factor of

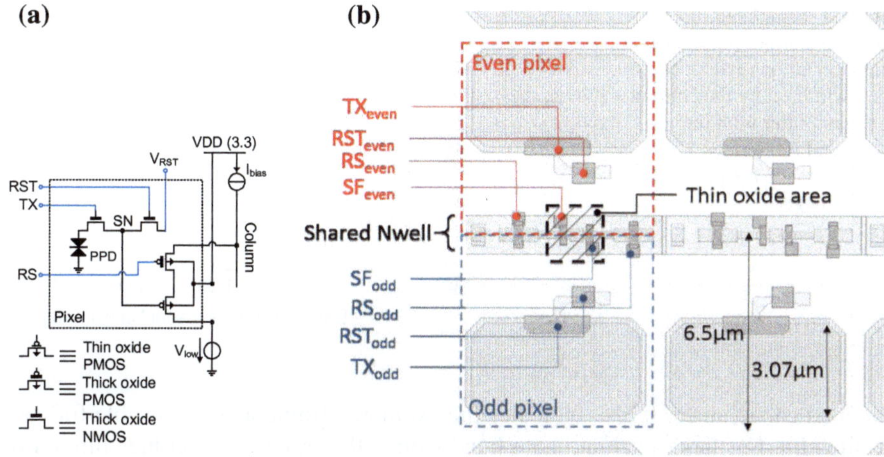

Fig. 7.2 Schematic of the proposed pixel (**a**) and a layout view of the neighboring pixels (**b**)

40% with a pixel pitch of 6.5 µm compared to a pitch of 6.5 µm in the chip presented in the last chapter.

7.1.3 Column-Level Amplifier

The noise reduction starts by implementing enough column-level gain in order to minimize the noise contribution of the next stages, limiting consequently the noise analysis and optimization to the in-pixel SF stage, the current source of the SF and the column-level amplifier.

The thermal noise originating from the SF stage and column-level amplifier is reduced using a proper design of the SF current source and bandwidth control obtained with high column-level amplification [1]. Simulation results from [2] show that thermal noise originating from the SF stage, the column-level amplifier and the current source can be reduced to about 0.2 e^-_{rms} if the bandwidth is limited to 265 kHz. Thermal noise can also be reduced using CMS at the end of the readout chain [3–5] but at the cost of additional circuitry when implemented in the analog domain and multiple analog-to-digital conversions when implemented in the digital domain. It has been shown in [1] that the input-referred thermal noise can also be reduced by increasing the conversion gain of the pixel. In this work we chose to perform a simple CDS for a faster readout. The bandwidth of the readout chain is roughly inversely proportional to the product of the column level gain and the load capacitance of the column level amplifier [1]. It has been set to 265 kHz with a gain of 64 in order to make the thermal noise lower than the $1/f$ noise (below 0.3 e^-_{rms}). Figure 7.3 shows the simulated input referred thermal noise of the readout chain as a function of its bandwidth which is tuned using different combinations of the column-level gain and

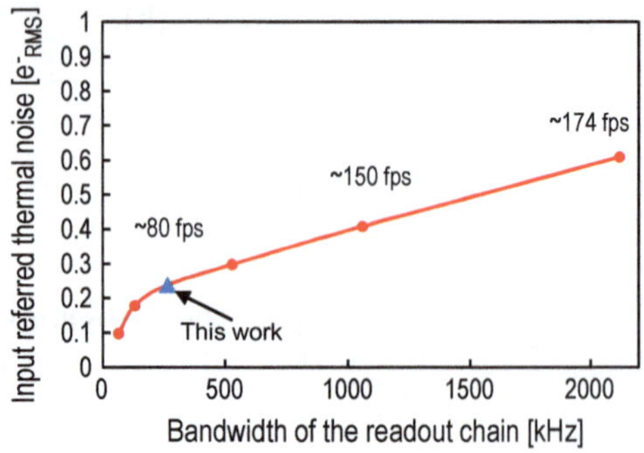

Fig. 7.3 Simulated input referred thermal noise as a function of the readout chain bandwidth set by the column level amplifier with the estimated framerates achievable for the presented VGA imager

the load capacitance. It also indicates the estimated framerates achievable for each bandwidth. For large readout chain bandwidths, the framerate becomes limited by the time required by the ADC to convert and shift the data to the output which is about 10 μs. At about 80 fps, the low frequency noise of the readout chain becomes the dominant noise source.

The $1/f$ noise is dramatically reduced by the AZ and CDS or CMS. However, it remains the dominant noise in the readout chain [2, 3]. The critical transistors of the column-level circuitry can be designed to have gate sizes large enough to make their contribution to the total $1/f$ noise of the readout chain negligible compared to the in-pixel SF's. Thus, in a conventional low noise CIS readout chain, with column amplification, bandwidth control and careful design, the $1/f$ noise originating from the in-pixel SF remains the dominant noise source.

Figure 7.4 shows the schematic of the column-level adjustable gain amplifier. The closed-loop gain is set by the ratio of the integrating and feedback capacitors. The feedback capacitors can be switched in order to change the gain between two levels: ×1, for high dynamic range in normal light conditions and ×64 for low light conditions. The open-loop gain is provided by an OTA. For the OTA design and layout, the priority is given to the noise constraint. The dynamic range is not critical since the voltage swing at the output of the pixel is not higher than 1.5 V. A single-ended structure is used because it involves half the number of noisy transistors compared to a differential one. Differential amplifiers are certainly better at rejecting any common mode noise (e.g. noise coming from the substrate and power supply) but at the cost of more noise and more power. Indeed, in the case of an operational transconductance amplifier (OTA) such as the one used in this amplifier, the differential structure requires to duplicate the circuit branch resulting in twice the power consumption and twice the thermal noise excess factor for achieving the same transconductance. In order to achieve low noise and to stay within our power budget, we have chosen a single ended implementation. The noise originating from the power and bias sources is reduced on-board by using power filters. Regarding the area, it is mainly set by the input and feedback capacitors especially for high gains. Hence, a differential topology would occupy about the same area but, as explained above, with the penalty of double power and noise. The situation would be even worse for a fully differential

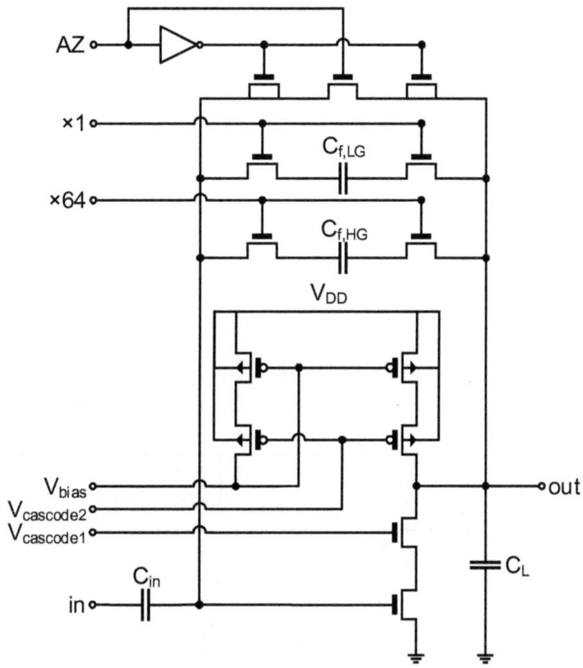

Fig. 7.4 Schematic of the column-level amplifier

amplifier (differential input differential output) since it would generate twice as much noise, power and area compared to the single ended implemented in this work. The high closed-loop gain of 64 requires a large open-loop gain hard to achieve with a simple single ended amplifier. It is known that cascode structures provide much higher gain with a negligible noise contribution of the cascode transistors. Hence, a fully cascoded single-ended amplifier is used. In order to make the $1/f$ noise contribution of the column-level amplifier negligible compared to the one originating from the pixel, the transistors of the OTA have gate areas more than 10 times larger than the SF. The charge injection of the AZ switch can reduce considerably the output voltage swing of the amplifier in the high column-level gain mode. In order to reduce that charge injection, dummy devices with a proper sizing are used in order to compensate the charge injected by the main NMOS switch.

7.1.4 Column-Level SSADC

Figure 7.5 shows the schematic of two neighbouring column-level SS-ADCs. A double stage comparator is used to reduce the offset [6]. Figure 7.6 shows the timing diagram of the whole readout chain. After the reset of the SN, the auto-zeros of the consecutive column-level amplifier and comparators are opened sequentially in order to minimize the impact of charge injection [7]. Then the charges are transferred

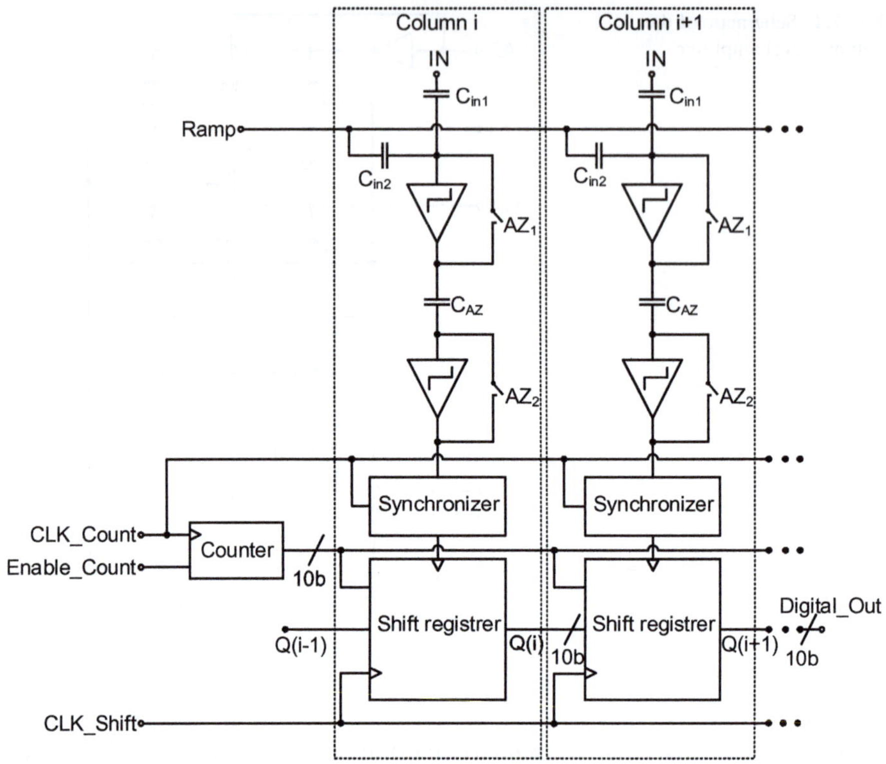

Fig. 7.5 Block diagram of the SS-ADC of two neighbouring columns

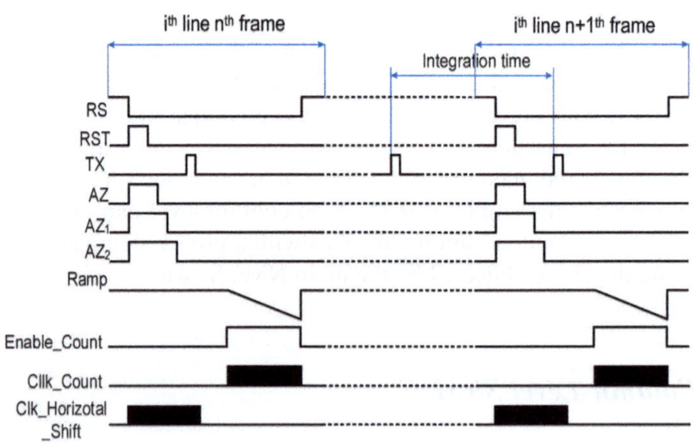

Fig. 7.6 Timing diagram of the proposed readout chain

to the SN and the voltage at the input of the comparator is equal to the difference between the transfer and reset levels. The ramp is then activated together with the counter. Shift registers are used in order to memorise the 10 bit code once the output of the comparators is high. The CDS time is defined by the time between the opening of the AZs and the moment when the output of the comparator is high. During the readout of the next line, the 10 bit codes of all the column-level registers are shifted horizontally to the digital output.

7.1.5 *Physical Implementation*

The chip has been fabricated in the same process than the pixels presented in Chap. 6, namely, a 180 nm CIS process with 4 metal layers, resulting in a chip area of 5 mm × 5 mm. The analog parts of the chip are powered with a 3.3 V source and digital parts with a 1.8 V. Figure 7.7 shows a chip micrograph locating the main design blocks. The VGA pixels array meant for a front side illumination occupies an area of 4.16 mm × 3.12 mm. Standard PPDs were used in the pixels array and the area between the even and odd pixels (Fig. 7.2b) occupied by the in-pixel transistors is covered by a metal layer (Fig. 7.8).

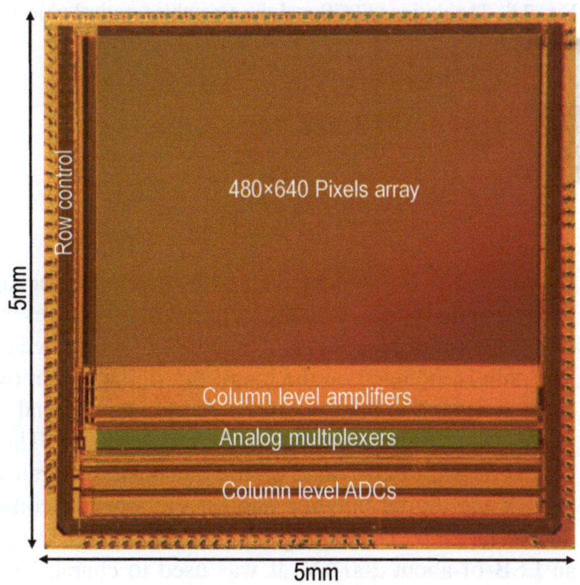

Fig. 7.7 Chip micrograph showing the main design blocks of the imager

Fig. 7.8 The designed PCB and the assembled optical objective used to test the presented image sensor

7.2 Test and Characterization

7.2.1 Experimental Setup

In order to test the presented chip, the latter was packaged in a PGA 144. A PCB has been designed to generate the different bias voltages and power supply ranging from 0 to 3.3 V, control the shape of the (TX) signal and generate the ramp of the SS-ADC. The chip is mounted on the board through a socket over which an optical objective is assembled. The board is powered with an external ± 5 V source and encompasses power supply filters for low noise requirements. The board is connected to a PXI rack with two FPGAs that, on the one hand, generate the digital control signals and receive the digital outputs, and on the other to convert the analog signal coming out of the chip. The analog input of the FPGA has an integrated 14 bit ADC with an LSB of about 200 μV. It was used to characterise the pixels using the analog output mode. The FPGAs are programmed with a computer using LABVIEW©. For measurements requiring the variation of the input light, a simple led powered with a low noise tunable voltage source was used. The LED was fixed at the end of a dark tube put on top of the imager.

7.2.2 Conversion Gain Measurement

In order to measure the conversion gain of the readout chain, the photon transfer curve (PTC) measurement technique is used [8]. Figure 7.9a shows the variance versus the signal PTC of the sensor measured from 5000 pixels. It is obtained using the analog output of the sensor and the column-level gain set to 64. The sensor is exposed uniformly to different levels of light provided by the LED and controlled with the voltage source. The variance and mean values are extracted from 100 images for each light level. The PTC curve collapses to zero around 2 V due to the column-level readout chain saturation. The measured overall conversion gain of the readout chain including the pixel, the column-level gain set to 64, the column-level analog buffer and the output analog buffer is measured by estimating the slope in the linear part of the PTC, which leads to 10.5 mV/e$^-$. Then, the pixel conversion gain is calculated by dividing the overall conversion gain by the gain of the column-level amplifier and the analog buffers which results in 160 μV/e$^-$. Note that the absolute value of the pixel conversion gain is not used for the input-referred noise measurement. Indeed the output noise RMS measured in mV at the output is directly divided by

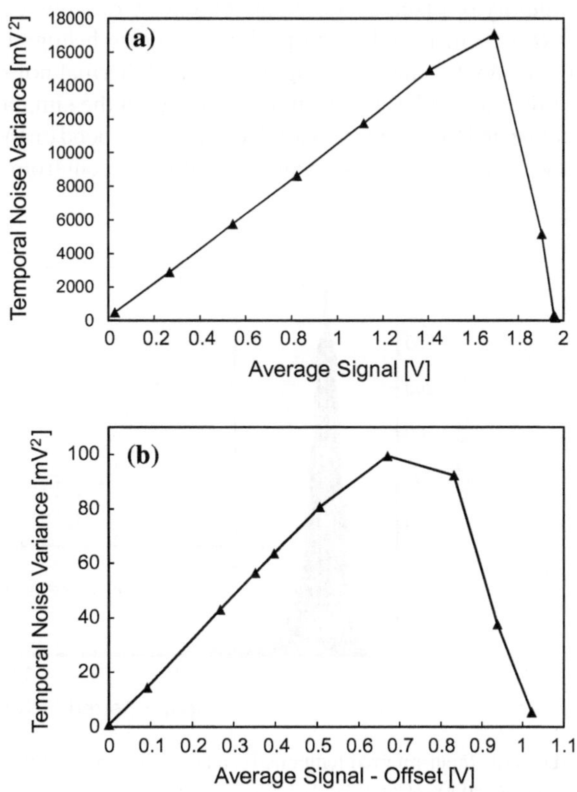

Fig. 7.9 Variance versus signal photon transfer curve (PTC) of the image sensor with **a** 64 column-level gain and **b** 1 column-level gain

the overall conversion gain of 10.5 mV/e$^-$. Figure 7.9b shows the PTC obtained by performing the same measurement with the column-level gain is set to 1. In this case, the PTC goes to zero around 1.02 V. This time, the saturation originates from the pixel. This curve allows the estimation of the pixel full well capacity. Using the measured conversion gain of 160 μV/e$^-$ the pixel full well capacity is about 6400 e$^-$.

7.2.3 Temporal Read Noise

In order to measure the input-referred noise of the presented imager, the output voltage noise is first measured. Then it is referred to the input using the readout chain conversion gain. This operation was applied to 5000 pixels after performing 100 readouts with a CDS of 5 μs and a line (pixel) readout time of 25 μs. The in-pixel SF current bias is set to 1.5 μA and the TX is off. The column-level amplifier gain is set to 64, limiting the bandwidth to about 300 kHz in order to reduce the thermal noise and the noise originating from the next stages. Figure 7.10 shows the resulting histogram of the input-referred TRN. The maximum of the histogram corresponds to 0.48 e^-_{rms}. The inset of Fig. 7.10 shows the histogram in a log scale highlighting a minority of pixels featuring an RTS noise of a few e^-_{rms}. The input referred noise of 0.48 e^-_{rms} measured at the peak of the noise histogram represents the total noise of the readout chain including the $1/f$ and thermal noise. Additionally, the estimation of the input referred $1/f$ noise is based on the simulated values of the capacitances connected to the sense node. These values depend on the layout and the process, hence it was expected to obtain values slightly different from the calculation and simulation.

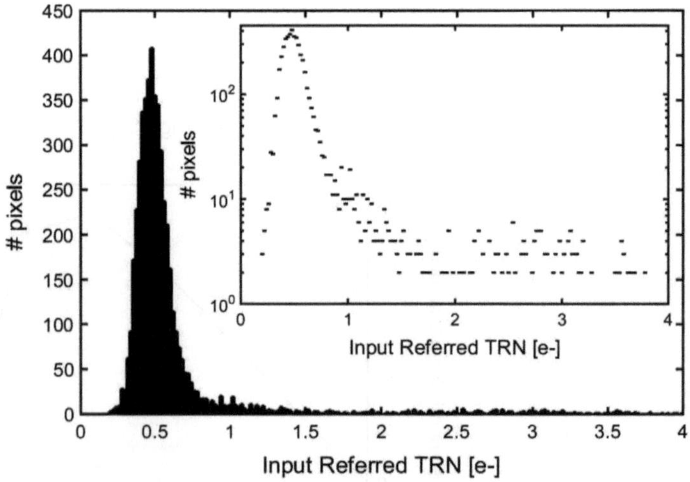

Fig. 7.10 Input-referred temporal read noise histogram of the image sensor with the vertical axis in linear and log scale (inset)

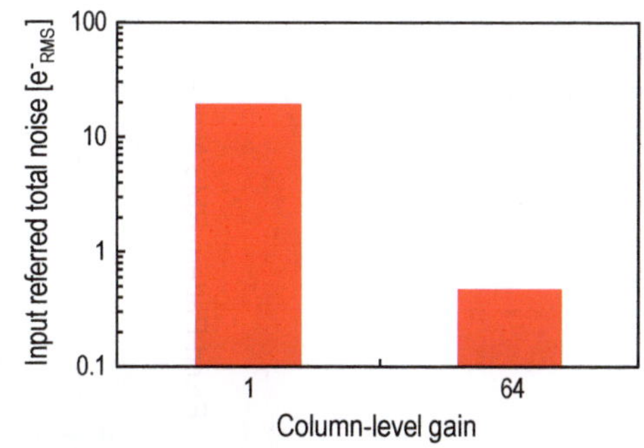

Fig. 7.11 Measured input referred noise in the log scale for the two column level gains

Figure 7.11 shows the input referred noise for the two column level gains. It shows a large noise increase in the $\times 1$ gain configuration as the bandwidth becomes much larger (more thermal noise) and all the noise sources after the column-level amplifier are no longer negligible. Thanks to the high column level gain of $\times 64$, the noise originating from the output buffers and analog to digital conversion is completely negligible.

7.2.4 *Photo-Response Non-uniformity*

This measurement is important in order to verify that the input-referred noise reduction does not come at the cost of higher photo-response non-uniformity (PRNU). The PRNU represents the spatial variation of the gain. It is given as an RMS percentage. For the presented imager, the PRNU was also measured using the PTC. The optical objective of the imager is removed. The chip is exposed directly to the LED put far enough to have a uniform illuminance for all the pixels array. 5000 pixels exposed to the same level of light with the same exposure time are read 100 times. The average output signal is calculated for each of the 5000 pixels. Then the standard deviation of the spatial variation of the pixels output voltage is plotted in Fig. 7.12 as a function of the average over time and space of the pixels output signal for different lighting conditions. The curve is linear as expected and the resulting PRNU corresponds to a value of 0.77%.

7.2.5 *Dark Current*

To measure the dark current, the imager is covered protecting it from any light. The chip is set to low column-level gain mode ($\times 1$). The average output signal of the sensor is measured, at room temperature, for different exposure times. Figure 7.13 shows the curve obtained by plotting the measured average output signals of the

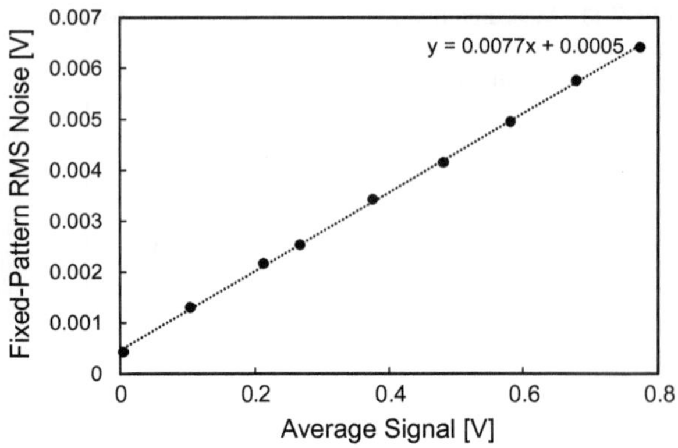

Fig. 7.12 PTC alike characterization of the fixed-pattern noise or PRNU. The measured value is 0.77%

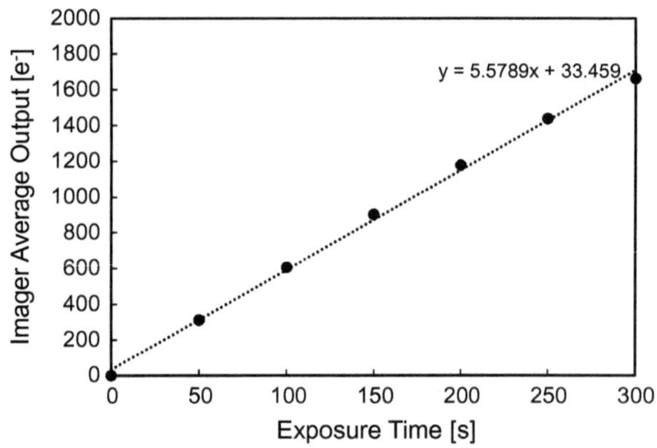

Fig. 7.13 Measurement of the dark current as the slope of the output versus exposure time curve

chip versus the exposure time. For integration times below 300 s, the number of accumulated charges in dark increases linearly with the integration time. The slope factor of the curve indicates a dark current of 5.6 e^-/s. Usually, integration times above a few tens of ms are not needed. Based on these measurements, the integrated charge originating from the dark current is not supposed to be more than a few 0.05 e^- which can still be neglected compared to the read noise.

7.2.6 Imager Lag

In order to measure the lag, the imager is exposed to a LED with a constant illumination. The imager exposure time is set to 50 ms. After this integration time, each pixel is read 2 times with a step of 20 μs. The signal integrated during 20 μs is considered negligible compared to the signal integrated during 50 ms and represents the floor for the lag assessment. The signal from the second readout of the pixel comes from the electrons left in the PPD after the charge transfer in the first readout. The

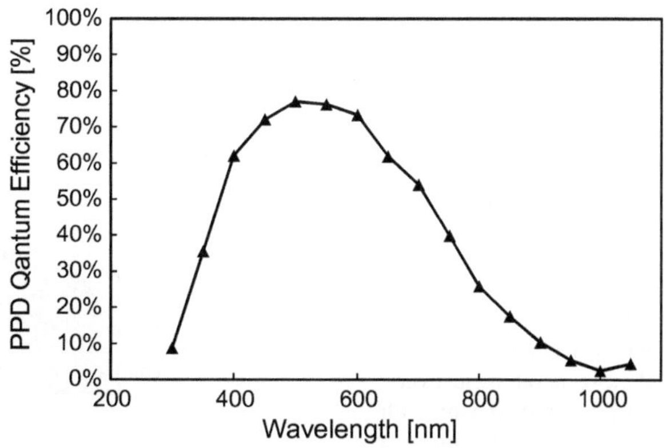

Fig. 7.14 Quantum efficiency of the PPD (active area) obtained by dividing the QE of the sensor by the fill factor of 40%

lag is obtained by dividing the signal from second readout over the one from the first readout. The lag was measured with different input light levels. When the PPD accumulates about 2500 photo-electrons the average measured lag was 0.1%. For an average number of accumulated electrons around 5000 the obtained lag increases to about 1%.

7.2.7 Quantum Efficiency

It is important to verify that the n-well containing the PMOS transistors of the pixel does not act as a photodiode competing with the PPD. To this purpose, a measurement of the quantum efficiency (QE) is required. Figure 7.14 shows the measured QE of the active area of the imager chip. It is obtained by dividing the effective QE by the pixel fill-factor of 40%. The resulting QE is as good as state-of-the-art PPDs designed exclusively using NMOS pixels. This means that the in-pixel n-well does not have a major impact on the PPD collection. Furthermore, the micro-lens layer has not been used in this imager. This layer focuses the light at the active area of the pixels and increases the effective QE.

7.2.8 Imaging Demonstration

Based on the measured performance, the presented imager is supposed to operate in both low light and normal light conditions. In order to validate this idea, images were taken using the presented image sensor in a dark room with a controlled level of light. The imager was set with an exposure time of 12 ms and a pixel readout time of 25 μs. Figure 7.15a and b show images taken with the chip under different low light levels using the ×64 column gain mode at an average of 3 and 41

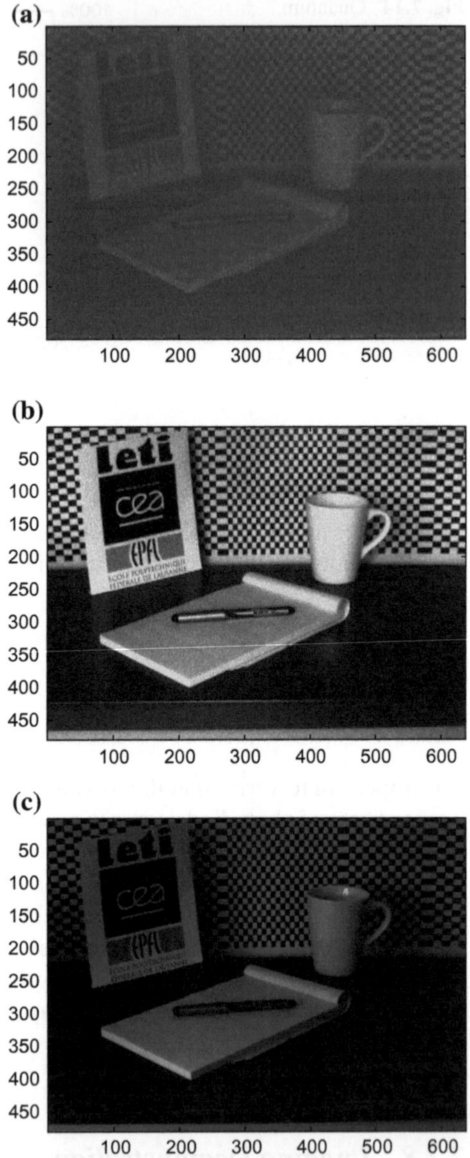

Fig. 7.15 Images, of the same scene, taken with the presented chip for different amounts of input light. **a** Shows the obtained image at 0.005 lux with a column-level gain set to 64 and an average of 3 photogenerated electrons per pixel. **b** At 0.066 lux with the column-level gain set to 64 and an average of 41 photo-generated electrons per pixel. **c** At 0.5 lux with the column-level gain set to 1 and an average of 320 photo-generated electrons

photo-generated electrons per pixel, respectively. In Fig. 7.15a, the SNR of the input light (limited by the photon shot noise) is below 5 dB. The dominant noise sources are the photon shot noise and the temporal read noise. One can see that even with such a low input light, the objects in the scene can still be distinguished. In Fig. 7.15b, the SNR of the input light is about 16 dB. Due to the high column gain, one can see that the image is not far from saturation due to voltage swing of the column-level amplifier. In these conditions, the photon shot noise and fixed pattern noise are supposed to be the dominant noise sources. Figure 7.15c shows an image obtained with day-light conditions at an average of 320 photo-generated electrons per pixel and the column-level gain set to 1. Here the input SNR is about 25 dB.

7.3 Discussion and Comparison to State-of-the-Art

The measurement results of the proposed imager, presented in the previous Section, are summarized and compared to recent state-of-the-art works in Table 7.1 showing that the proposed noise reduction technique is efficient. The noise is reduced, as expected, to a record low level of 0.48 e^-_{rms} in room temperature which is about 1.5 times lower than state-of-the-art [9, 10] in a full imager with the conventional SF based readout scheme. This record low noise does not come at the cost of a slow readout, a low fill factor or a low SNR. Indeed the line readout time of 25 μs is actually 64 and 5.7 times faster than [9, 10], respectively. A dynamic range of 82.5 dB in the dual gain mode can be achieved.

The improvements achieved in this work were obtained only by making design choices and careful layout based on a detailed noise analysis [1]. The thin oxide transistor used as an SF is a standard "digital" transistor not optimized for analog applications, thus there is still some room left for process optimization at the SF transistor level. Moreover, (4.28) suggests that the input-referred noise can also be mitigated by reducing the contribution of all the parasitic, overlaps and junction capacitances to the sense node (C_P). But this approach requires process refinements. Indeed, in [10], the overlap capacitances were reduced by using reset, transfer and gates without low doped drains. In a recently reported work, a mean noise of 0.29 e^-_{rms} was measured on a small array of pixels (12 × 12), by using a low doped PN junction isolated from the transfer gate as well as a tapered SF [11]. This result came at the cost of a low full well capacity and a relatively higher lag. In [12], 0.27 e^-_{rms} at −10°C has been reported by implementing a special implant isolating the FD from the TG (reducing the overlap capacitance) and by replacing the reset transistor by an implant connected to a reset clock. But this was achieved at the cost of a high off-chip reset voltage of 25 V and a low full well capacity of 1500 e^-. No characterization has been reported in order to verify the impact of these process modifications on the global performance of the pixels. Note that these process level techniques are compatible with the circuit techniques presented in this work. A combination between the reduction of C_P and using thin oxide SF with optimized gate size is expected to come with even more noise reduction. Indeed, the calculation results presented in Fig. 5.10 show that the input referred noise can be reduced to 0.29 e^-_{rms} if the capacitance C_P of the sense node is reduced to 0.25 fF, using process refinements [13], instead of the 0.75 fF simulated in the standard process.

Table 7.1 Summary of the imager performance with comparison to state of the art

Reference	This work [14]	[9]	[10]	[11]	[12]
Circuit level noise reduction techniques	Standard thin oxide PMOS SF and column-level gain	Column-level gain and CMS	Column-level gain	Column-level gain and off-chip CMS	Column-level gain and CMS
Process modifications for noise reduction	Standard process	Buried channel SF	Buried channel SF and process refinements to reduce the SN capacitance	Special potential profile for a FD distant from the TG and tapered reset transistor	Special implants for distant FD from the TG and replacement of the reset transistor by an implant connected to a 25 V reset clock voltage
Process	180 nm CIS	180 nm CIS	180 nm CIS	BSI[a] CIS	110 nm CIS
Active array size	640(H)×480(V)	128(H)×198(V)	180(H)×480(V)	12(H)×12(V)	25(H)×512(V)
Pixel size [μm^2]	6.5×6.5	10×10	5.5×5.5	1.4×1.4	11.2×5.6
Fill factor [%]	40	33	48.8	N.R	N.R
Conversion gain [$\mu V/e^-$]	160	45	240	413.4	220
Full well capacity [e^-]	6400	N.R	76000[b]	210	1500
Column-level gain	×1 ×64	×1 ×4 ×16	Dual gain	×8 ×16 ×24	×128
Pixel readout time [μs]	25	1600	143	N.R	N.R
Read noise (peak) [e^-_{rms}]	0.48 @ room $T°$	0.7 @ room $T°$	0.74[c] @ room $T°$	0.29 @ room $T°$	0.27 @ −10°
Dark current [e^-/s]	5.6 @ room $T°$	N.R	N.R	0.1 @ room $T°$	N.R
PRNU [%]	0.77	N.R	N.R	N.R	N.R
Imager lag [%]	0.1	N.R	N.R	1	N.R

[a] Back Side Illumination
[b] Using a lateral overflow capacitance
[c] The noise reported when the pixel is operated in the source follower configuration

7.4 Conclusion

This Chapter demonstrates that sub-0.5 electron noise, in a full VGA APS, can be achieved using a standard CIS process by a proper circuit noise optimization exploiting all the degrees of freedom left to the designer for minimizing the total input-referred noise. The proposed techniques include the following steps: (a) introduce enough column level gain to, on one hand, reduce the pixel and column-level amplifier thermal noise and on the other the noise contribution of the next stages, (b) design the column-level amplifier with a minimum number of devices large enough to make the residual noise after auto-zero small enough compared to the SF noise, (c) replace the conventional thick oxide NMOS SF with a minimum width and optimum length thin oxide PMOS, (d) draw a compact layout with common n-well and minimum sense node parasitic capacitance. All these measures result in the majority of the full VGA array pixels peaking at an input-referred TRN of $0.48\,e^-_{rms}$ and a minority of pixels showing an RTS noise of a few electrons. The input-referred TRN is measured without CMS and at a short pixel (line) readout time, for a sub-electron read noise CIS, of 25 μs.

This work also provides a full characterization of the VGA imager showing that neither of the dynamic range, the imager lag nor the PRNU are compromised with the proposed noise reduction technique. The characterization also shows that the QE of the PPD is not affected by the neighbouring PMOS n-wells.

Note that the proposed approach can be combined with any known additional noise reduction techniques at system (e.g. CMS), device (e.g. buried channel) and process level (e.g. C_P reduction) to further reduce the TRN.

References

1. A. Boukhayma, A. Peizerat, C. Enz, Temporal readout noise analysis and reduction techniques for low-light cmos image sensors. IEEE Trans. Electron Devices **63**(1), 72–78 (2016)
2. A. Boukhayma, A. Peizerat, A. Dupret, C. Enz, Design optimization for low light cmos image sensors readout chain, in *2014 IEEE 12th International New Circuits and Systems Conference (NEWCAS)* (2014), pp. 241–244
3. S. Suh, S. Itoh, S. Aoyama, S. Kawahito, Column-parallel correlated multiple sampling circuits for cmos image sensors and their noise reduction effects. Sensors **10**(10), 9139–9154 (2010)
4. A. Boukhayma, A. Peizerat, C. Enz, A correlated multiple sampling passive switched capacitor circuit for low light cmos image sensors, in *2015 International Conference on Noise and Fluctuations (ICNF)* (2015), pp. 1–4
5. C. Enz, A. Boukhayma, Recent trends in low-frequency noise reduction techniques for integrated circuits, in *2015 International Conference on Noise and Fluctuations (ICNF)* (2015), pp. 1–6
6. R. Poujois, J. Borel, A low drift fully integrated mosfet operational amplifier. IEEE J. Solid-State Circuits **13**(4), 499–503 (1978)
7. C. Enz, G. Temes, Circuit techniques for reducing the effects of op-amp imperfections: autozeroing, correlated double sampling, and chopper stabilization. Proc. IEEE **84**(11), 1584–1614 (1996)
8. J.R. Janesick, *Photon Transfer* (SPIE, USA, 2007), pp. 49–79

9. Y. Chen, Y. Xu, Y. Chae, A. Mierop, X. Wang, A. Theuwissen, A 0.7e-rms temporal-readout-noise cmos image sensor for low-light-level imaging, in *2012 IEEE International Solid-State Circuits Conference Digest of Technical Papers (ISSCC)* (2012), pp. 384 –386
10. S. Wakashima, F. Kusuhara, R. Kuroda, S. Sugawa, A linear response single exposure cmos image sensor with 0.5e-readout noise and 76ke- full well capacity, in *2015 Symposium on VLSI Circuits (VLSI Circuits)* (2015), pp. C88–C89
11. J. Ma, D. Starkey, A. Rao, K. Odame, E. Fossum, Characterization of quanta image sensor pump-gate jots with deep sub-electron read noise. IEEE J. Electron Devices Soc. **3**(6), 472–480 (2015)
12. M. Seo, S. Kawahito, K. Kagawa, K. Yasutomi, A 0.27e^-_{rms} read noise 220$\mu v/e^-$ conversion gain reset-gate-less cmos image sensor with 0.11- μm cis process. IEEE Electron Device Lett. **36**(12), 1344–1347 (2015)
13. K. Fumiaki, W. Shunichi, N. Satoshi, K. Rihito, S. Shigetoshi, Analysis and reduction of floating diffusion capacitance components of cmos image sensor for photon-countable sensitivity, in *International Image Sensors Workshop* (2015)
14. A. Boukhayma, A. Peizerat, C. Enz, A sub-0.5 electron read noise vga image sensor in a standard cmos process. IEEE J. Solid-State Circuits **51**(9), 2180–2191 (2016)

Chapter 8
A Passive Switched-Capacitor Circuit for Correlated Multiple Sampling

8.1 Introduction

It has been shown, in the previous chapters, that for CIS readout chains based on pinned photodiodes, the combined $1/f$ and thermal readout circuit noise can be reduced using CMS. In the last few years, CMS has been implemented at column level for more efficient $1/f$ and thermal noise reduction. Two main implementations have been introduced, one using additional column level active circuitry to integrate multiple samples in one capacitor [1] and the other one performs the CMS after the analog-to-digital conversion using multiple conversions [2].

8.2 Motivation

The analog implementation of a CMS of order M consists in using a SC amplifier that accumulates M consecutive samples in its feedback capacitor [1]. The drawback of this technique is that accumulation of voltages reduces the dynamic range of the readout chain by a factor of M. An analog alternative is the folding integration technique [1]. It allows performing CMS without reducing the dynamic range at the cost of adding two voltage references, one comparator, and some control logics. This additional circuitry introduces a feedback effect in order to prevent the output of the SC amplifier from entering in saturation.

The implementation of CMS in the digital domain consists in the addition of the first M samples (reset) after analog-to-digital conversion and then subtract consecutively the next M ones (transfer) [2–4]. The main drawback of this technique is the fact that analog-to-digital conversion has to be performed $2M$ times during each readout. The motivation of this chapter is to propose a CMS analog implementation that (i) requires no additional active circuitry to preserve power consumption, (ii) has no impact on the output dynamic range and (iii) does not need multiple analog-to-digital conversions.

A. Boukhayma, *Ultra Low Noise CMOS Image Sensors*,
Springer Theses, https://doi.org/10.1007/978-3-319-68774-2_8

8.3 Design Principle

The alternative CMS implementation proposed in this chapter relies on analog averaging using switched capacitors (SC). It is based on a passive SC network using an optimal number of capacitors. Let us consider two capacitors, C_1 and C_2, each one holding, respectively, voltages V_1 and V_2 as shown in Fig. 8.1. When connecting both capacitors by closing the switch S_1, the charges held in C_1 and C_2 are shared. The charge conservation leads to a common voltage V across both capacitors given by

$$V = \frac{C_1 V_1 + C_2 V_2}{C_1 + C_2} \overset{C_1 = C_2}{=} \frac{V_1 + V_2}{2} \tag{8.1}$$

The simplest way to perform an average of M samples using switched capacitors consists of holding M consecutive samples with a period of sampling T_S in M capacitors and connecting them all, at $M \cdot T_S$, to obtain the average value of the M samples. This process is faster than the state-of-the-art techniques reported in the previous section at the cost of the silicon area occupied by the M sampling capacitors. In order to use less capacitors to average the same number of samples, one has to process the average progressively and hold the intermediate results in the capacitors instead of holding the initial samples. Consider the case where $M = 2^n$. One can find that the average of the 2^n consecutive samples can be calculated recursively. In fact the average of 2^n samples is calculated by connecting two capacitors, each one holding the average of 2^{n-1} consecutive samples as

$$\frac{1}{2^n}\sum_{i=1}^{2^n} V_i = \frac{1}{2}\left(\frac{1}{2^{n-1}}\sum_{i=1}^{2^{n-1}} V_i + \frac{1}{2^{n-1}}\sum_{i=2^{n-1}}^{2^n} V_i\right). \tag{8.2}$$

Figure 8.2 illustrates the progressive processing of the samples average over the time needed to sample and process the total 2^n samples: $(2^n T_S)$. Note that the signal settling time after connecting the capacitors is considered negligible compared to the sampling period T_S (8.2) and Fig. 8.2 show that, at $2^n T_S$, the average of 2^n samples is processed by connecting two capacitors, one storing the average of the first 2^{n-1} samples processed at $2^{n-1}T_S$, and the other one storing the average of the next 2^{n-1} samples. One can notice that the capacitors used to compute the average of the first 2^{n-1} samples are not all needed to hold this value during the processing of the next samples. Only one capacitor is needed to hold the average of the first 2^{n-1} samples and the other capacitors can be reused to process the next samples. Consequently, if we consider U_{n-1} to be the number of capacitors needed to calculate the average of 2^{n-1} consecutive samples. The average of 2^n samples can be calculated using $U_{n-1} + 1$ capacitors. Hence:

$$U_n = U_{n-1} + 1. \tag{8.3}$$

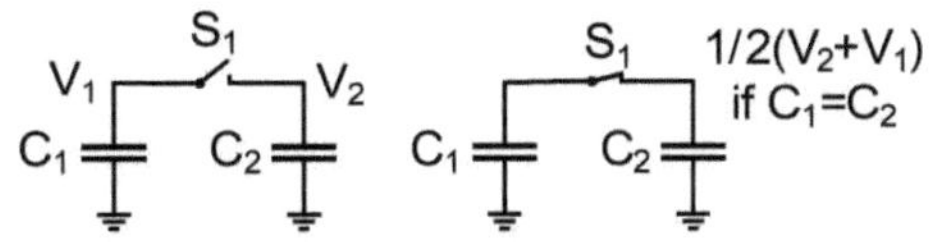

Fig. 8.1 Charge conservation in two switched capacitors leading to the mechanism of averaging

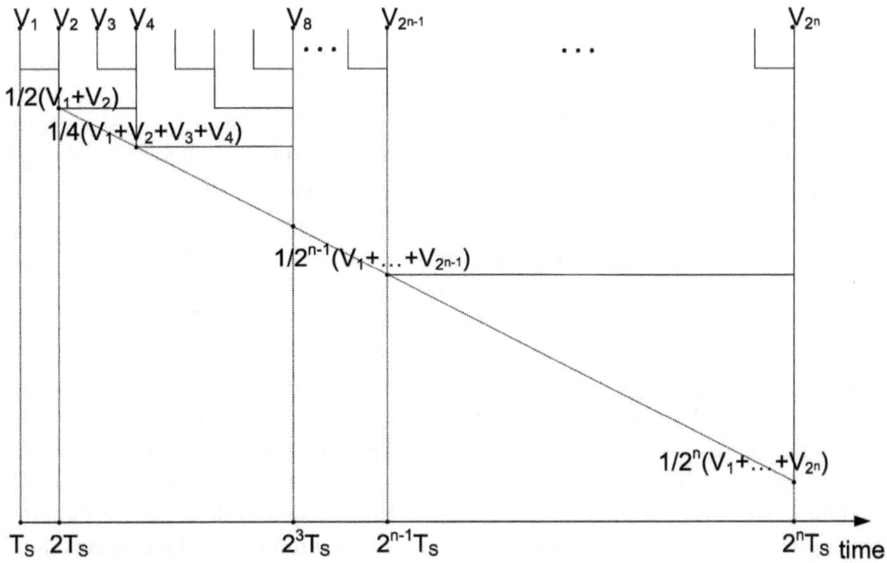

Fig. 8.2 Illustration of the progressive processing of the average of 2^n consecutive samples with a sampling frequency T_S

Based on this recursive relation (8.3), the number of capacitors needed to calculate the average of 2^n consecutive samples is given by

$$U_n = n + 1. \tag{8.4}$$

This result shows that a CMS of order 2^n can be performed using $n + 1$ capacitors instead of 2^n. Therefore, the implementation of CMS using this technique results in a CMS layout footprint reduction of order $\frac{2^n}{n}$.

8.4 Implementation

Figure 8.3 shows the schematic of the passive SC network calculating averages of order 2^n using n capacitors. Consider the case $n = 2$. Figure 8.4 shows the steps of the calculation of the average of 4 consecutive samples. First, switches S_1 and S_2 are closed. S_2 is opened to hold a sample V_1 in C_{out}. Then S_1 is opened after T_S in order to hold the next sample V_2 in C_1. S_1 is closed to calculate the average and opened to hold a voltage $\frac{1}{2}(V_1 + V_2)$ in C_{out}. The same process is iterated with capacitors C_1 and C_2 and ends with a voltage $\frac{1}{2}(V_3 + V_4)$ held in capacitor C_2. Finally S_4 is closed to connect C_2 and C_{out} and then opened to hold a voltage $\frac{1}{4}(V_1 + V_2 + V_3 + V_4)$ in C_{out}.

Figure 8.5 shows the implementation of the averaging circuit presented in Fig. 8.3 in a conventional CIS readout chain to perform CMS of order 2^n. Note that the input capacitors of the ADC comparator C_{comp1} and C_{comp2} are small compared to C_{out}. This is because the total capacitance seen at the output of the CMS stage is given by $C_{\text{out}} + C_{\text{comp1}}$ during the reset level averaging phase (the switch AZ_{comp} is opened) and $C_{\text{out}} + \frac{C_{\text{comp1}} \cdot C_{\text{comp2}}}{C_{\text{comp1}} + C_{\text{comp2}}}$ after the transfer (the switch AZ_{comp} is closed). The corresponding timing diagram for the case $n = 3$ is depicted in Fig. 8.6. The voltage at the input of the comparator corresponds to the difference between the voltage levels before and after opening the auto-zeroing switch (AZ). The same averaging circuit computes the average of 2^n samples during the reset phase and the average is stored in capacitor C_{comp1}. Then, the auto-zeroing switch (AZ) is opened and the voltage at the input of the comparator becomes the difference between the average current output and the average calculated at the reset phase. Once the average of the 2^n samples of the transfer phase is calculated, the ramp is activated together with the counter as shown in Fig. 8.6 for the case $n = 3$. Compared to a conventional CIS readout chain, only two additional capacitors are needed with no active circuitry. The logic circuit needed to control the switches is common to all the columns of the imager, thus its footprint is not significant.

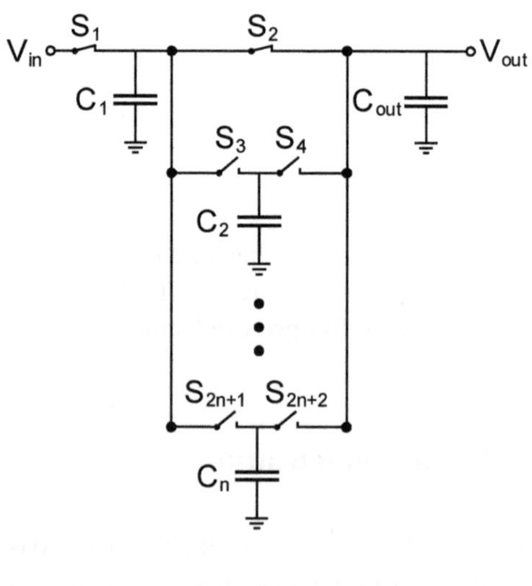

Fig. 8.3 Passive SC circuit averaging 2^n samples with n capacitors

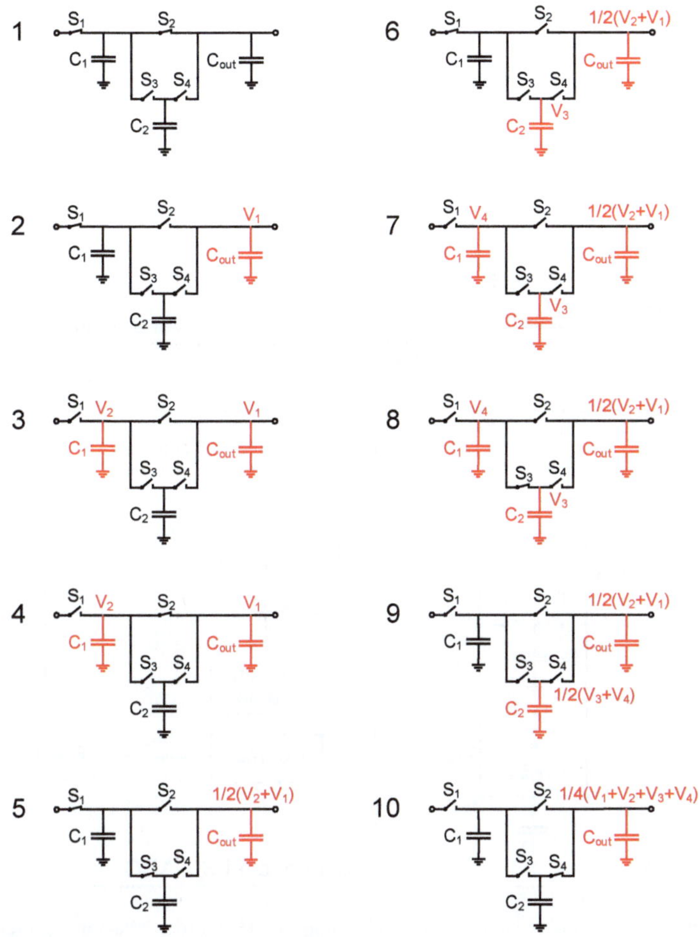

Fig. 8.4 Processing of the average of 4 samples with the SC network of Fig. 8.3

8.5 Transient Noise Simulations

In order to validate the proposed CMS circuit presented in this paper, the CIS readout chain presented in Fig. 8.5, using the analog averager for the case $n = 3$ to perform a CMS of order 8, is simulated using the ELDO © transient noise simulation. The line readout time is set to 65 µs for a column amplifier bandwidth of 265 kHz and gain of 8 (in order to minimize the contribution of the ADC comparator to the input referred noise). The design kit used for this simulation is a 180 nm process dedicated to CIS. Simulations of $1/f$ and thermal noise are performed separately. Note that the averaging circuit can perform CMS of orders from 1 (corresponding to the simple CDS) up to 8. Figure 8.7 shows the input-referred thermal noise as a

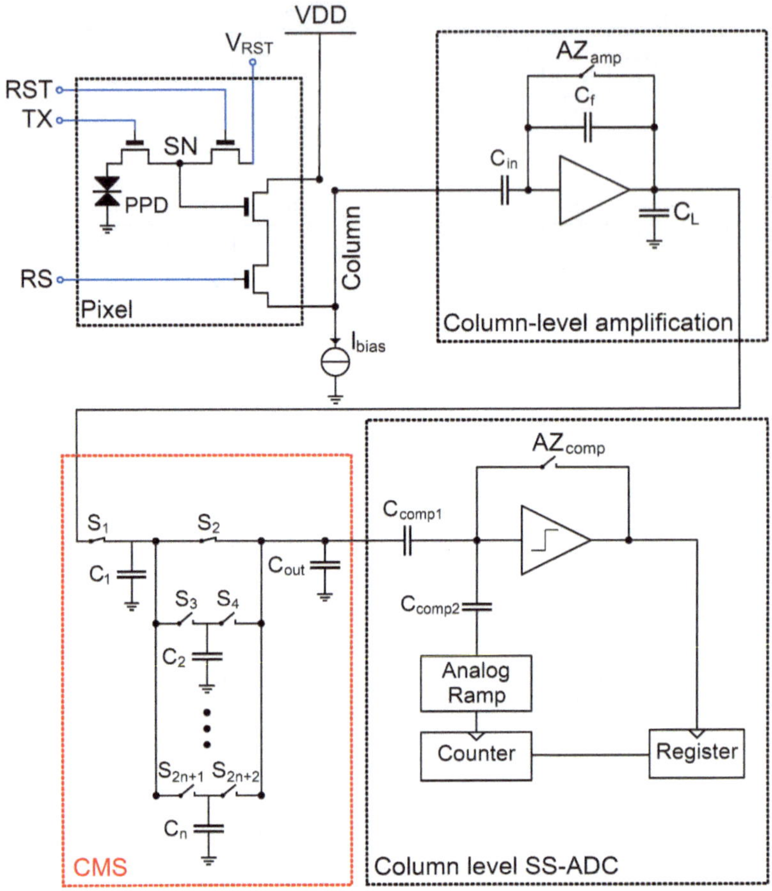

Fig. 8.5 Implementation of the new circuit performing a CMS of order 2^n using n switched capacitors in a conventional CIS readout chain

function of the CMS order M together with a $1/\sqrt{M}$ theoretical decrease expectation curve. This figure shows that, as expected, thermal noise reduction follows a $1/\sqrt{M}$ decrease. Figure 8.8 shows the $1/f$ component of the input referred noise for M varying between 1 and 8. It shows that CMS reduces $1/f$ noise by about 30% for a CMS of order 4 and 33% for order 8. Thus for $1/f$ noise using CMS of orders higher than 4 does not bring any significant decrease.

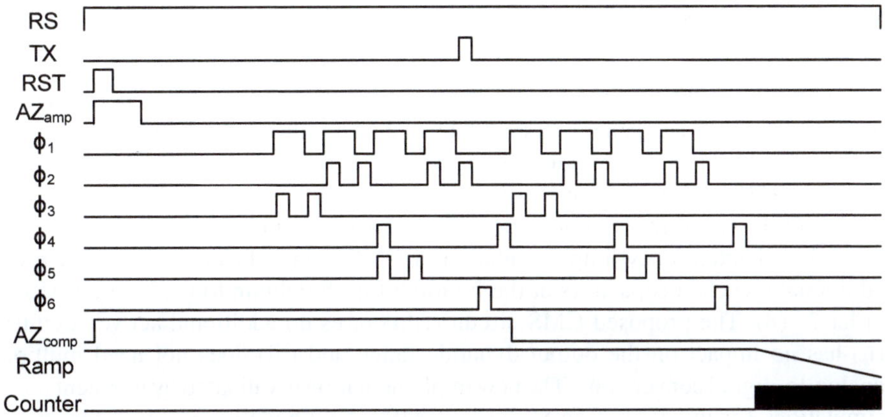

Fig. 8.6 Timing diagram for the CIS readout chain of Fig. 8.5 that uses the proposed CMS implementation for $n = 3$ (CMS of order 8)

Fig. 8.7 Simulations of the input-referred thermal noise rms charge of the CIS readout using the proposed CMS circuit

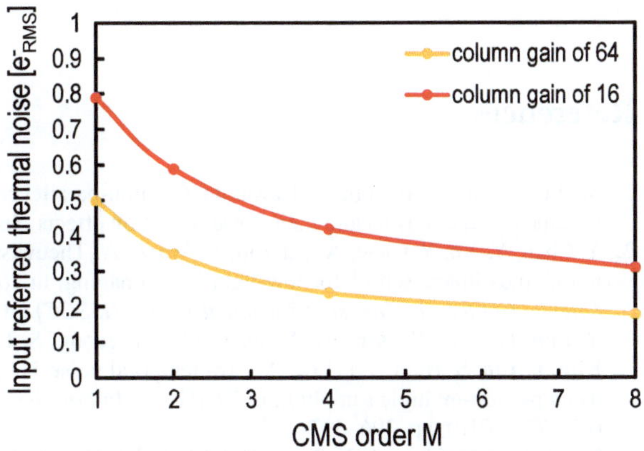

Fig. 8.8 Simulations of the input-referred $1/f$ noise rms charge of the CIS readout chain using the proposed CMS circuit

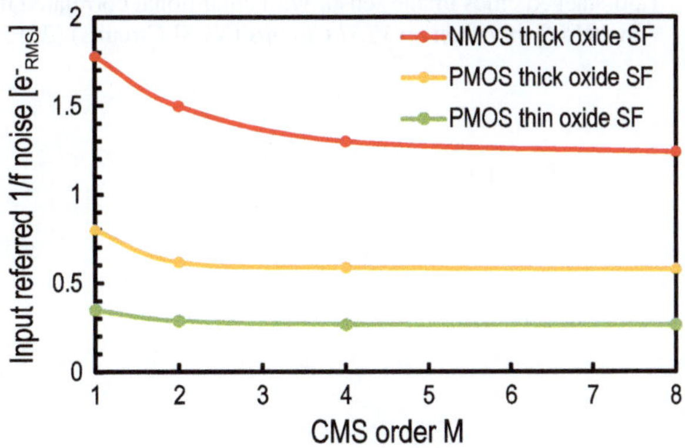

8.6 Conclusion

In a CIS readout chain based on a 4T pixel, a CMS of order M reduces thermal noise by $\frac{2}{M}$. CMS is a general case of CDS. Thus it reduces significantly $1/f$ noise and offset. The $1/f$ noise reduction using CMS increases with M and reaches a plateau at $M = 8$. In state-of-the-art CIS readout chains, CMS is performed after analog-to-digital conversion or using analog active circuitry.

In this chapter, a new implementation of CMS is introduced. It uses only n (3) additional switched capacitors at the bottom of each column to perform a CMS of order 2^n (8). The proposed CMS circuit (i) requires no additional active circuitry, (ii) has no impact on the output dynamic range and (iii) does not need multiple analog-to-digital conversions. The new implementation is validated by transient noise simulations for the case of an 8th order CMS. The simulation results confirm the theoretical results presented in Chap. 5. They show that the thermal noise is decreased as expected as $\frac{1}{\sqrt{M}}$, whereas the $1/f$ noise is reduced by 30% for $M = 4$ but no significant additional decrease occurs for M larger than 8.

References

1. S. Suh, S. Itoh, S. Aoyama, S. Kawahito, Column-parallel correlated multiple sampling circuits for cmos image sensors and their noise reduction effects. Sensors **10**(10), 9139–9154 (2010)
2. Y. Chen, Y. Xu, Y. Chae, A. Mierop, X. Wang, A. Theuwissen, A 0.7e-rms temporal-readout-noise cmos image sensor for low-light-level imaging, in *2012 IEEE International Solid-State Circuits Conference Digest of Technical Papers (ISSCC)* (2012), pp. 384–386
3. Y. Lim, K. Koh, K. Kim, H. Yang, J. Kim, Y. Jeong, S. Lee, H. Lee, S. H. Lim, Y. Han, J. Kim, J. Yun, S. Ham, Y.T. Lee, A 1.1e- temporal noise 1/3.2-inch 8mpixel cmos image sensor using pseudo-multiple sampling, in *2010 IEEE International Solid-State Circuits Conference - (ISSCC)* (2010), pp. 396–397
4. S.F. Yeh, K.Y. Chou, H.Y. Tu, C.Y.P. Chao, F.L.Hsueh, A 0.66e rms temporal-readout-noise 3d-stacked cmos image sensor with conditional correlated multiple sampling (ccms) technique in *2015 Symposium on VLSI Circuits* (VLSI Circuits) (2015), pp. C84–C85

Chapter 9
Downscaling Effects Towards Photon Counting Capability in CIS

The idea of an image sensor with photon counting capability is becoming a subject of interest for new applications and imaging paradigms like quanta image sensors [1, 2]. Such a device must have an input-referred read noise negligible compared to a single electron. Among the state-of-the-art imaging devices, single photon detectors may appear to be the best candidate for such an application [3]. Historically, microelectronics could not provide readout chains with noise levels as low as deep sub-electron. Hence, the solution was to introduce a gain at the level of the photon-electron conversion. In photomultipliers tubes (PMTs) and single photon avalanche photodiodes (SPADs) the electron generated by the incident photon is accelerated and multiplied to a number of electrons from a few hundreds in PMTs to millions in SPADs. Such a signal level can be easily detected and quantized into two logic levels since the number of incident photons during the period of detection is assumed to be much less than one. But these devices present the following disadvantages [4]. First, they are limited to the case of single photon detection. In other words, the arrival of one photon and multiple photons are not distinguished. Second, these devices suffer from a dead time and after pulse following each photon detection blinding the device for a certain time. The third limitation is related to the low resolution and fill factors of focal plane arrays using such devices. Additionally, they use high voltages which are not compliant with standard CIS processes.

During the last decade, CIS have seen their performance increasing remarkably in terms of dynamic range, speed, resolution and power consumption. With a lower cost and better on-chip integration, CIS replaced progressively the charge coupled devices (CCDs) in many applications and enlarged the market of electronic imaging devices. In terms of sensitivity, the quantum efficiency have been improved to reach levels as high as 95% [2]. The fill factors have been constantly improved. The dark

A. Boukhayma, *Ultra Low Noise CMOS Image Sensors*,
Springer Theses, https://doi.org/10.1007/978-3-319-68774-2_9

current in the PPDs have been reduced to levels making the process of electron-hole pair generation noiseless for integration times around tens of ms. The read noise has also been dramatically reduced to reach deep sub-electron levels [5–7]. Hence, CIS technologies have progressed enough now to eventually envisage photon counting.

9.1 Photon and Electron Count Conditions

For a semiconductor device, an impinging photon with an energy higher than its bandgap may generate or not a photoelectron. The efficiency of the photon absorption and generation of an electron hole pair is evaluated by the quantum efficiency (QE). The QE is measured as the average number of generated photoelectrons out of an average number of incident photons. In other words, it is measured by performing multiple shots in stationary conditions in order not to compromise the measured value by the shot noise ubiquitous with photons incidence.

The photoelectron generation is, hence, a stochastic process governed by a the binomial law and involves the quantum efficiency of the detection device [2]. The probability of generating k photoelectrons out of n incident photons is given by

$$p(k, n) = \frac{n!}{k!(n-k)!} \cdot QE^k \cdot (1 - QE)^{n-k}. \tag{9.1}$$

Hence, the probability of generating n photoelectrons out of n incident photons is given by QE^n. Consequently, the quantum efficiency is a very critical parameter for incident photons counting since $p(n, n)$ decreases exponentially with n for a QE exclusively smaller than 1.

Counting the photoelectrons corresponds to a quantization of the signal generated by these electrons. In the case of CIS, the signal corresponds to the voltage drop at the level of the sense node after transferring the photoelectrons from the PPD. Quantizing the voltage level with a step equal to the conversion gain of the CIS readout chain allows counting the photons ideally. But this conversion gain step has to be higher enough than the read noise standard deviation in order to perform accurate quantization and photoelectrons counting.

The probability of performing a true photoelectron count and single photoelectron detection can be calculated by assuming a Gaussian distribution of noise, centered at each quantized step, and using the error function [2]. Figure 9.1 shows the calculated probability of a true photo-electron count and a single photo-electron detection as a function of the input-referred readout chain noise. It shows that a 90% accuracy requires a read noise below $0.4\,e^-_{rms}$ for single photo-electron detection and $0.3\,e^-_{rms}$ for photo-electron count. Recently reported works are today closer than ever to these limits [6–8]. Here after, the possibility of performing sub-$0.3\,e^-_{rms}$ by taking advantage of the technology downscaling is discussed.

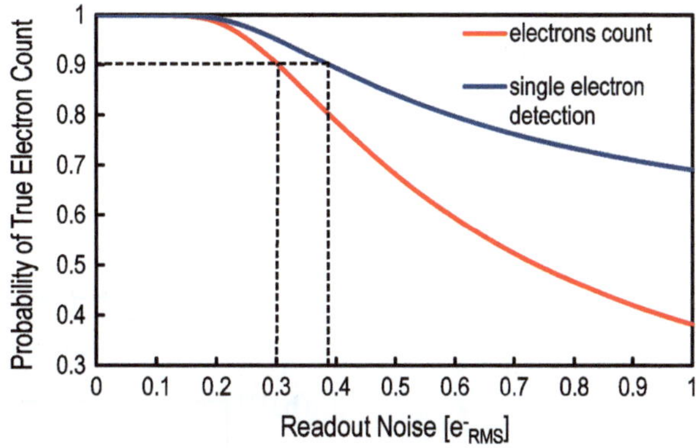

Fig. 9.1 Probability of a true photo-electron count and single photo-electron detection as a function of the input-referred readout noise [2]

9.2 Impact of Technology Downscaling on the Read Noise

9.2.1 1/f and Thermal Noise

Since their first development, CIS pixels have always been designed with thick oxide transistors compatible with high voltages (3.3 V). The device parameters of thick oxide transistors do not follow the scaling rules as the thin oxide transistors. The performance improvement of technology downscaling on these devices is rather limited. Moreover, it appears that the oxide trap density of thick oxide transistors tends to increase with the technology downscaling as shown in Fig. 9.2b.

It has been demonstrated, earlier in this work, that a thin oxide SF can be used together with a conventional PPD for a low noise performance. Thin oxide transistors on the other hand take full advantage of the technology downscaling. Thus, it is interesting to investigate the impact of the technology downscaling on the input-referred noise. The starting point for analyzing the impact of the technology downscaling on the input-referred noise of a readout chain based on a thin oxide SF is (4.28) and (4.24). The conclusions can be made based on how the technology downscaling affects the different process and device parameters. Table 9.1 shows the scaling factor corresponding to the relevant device parameters [9]. The technology downscaling allows a higher oxide capacitance per unit area, a lower gate width and lower overlap and parasitic capacitances. Hence, the input-referred $1/f$ noise variance is supposed

Table 9.1 Imapct of the technology downscaling on the parameters of the input-referred $1/f$ noise in (4.28)

Parameter	Scaling factor
W	$\frac{1}{\kappa}$
L	$\frac{1}{\kappa}$
C_{ox}	κ
C_P	$\frac{1}{\kappa}$
$C_e \cdot W$	$\frac{1}{\kappa}$

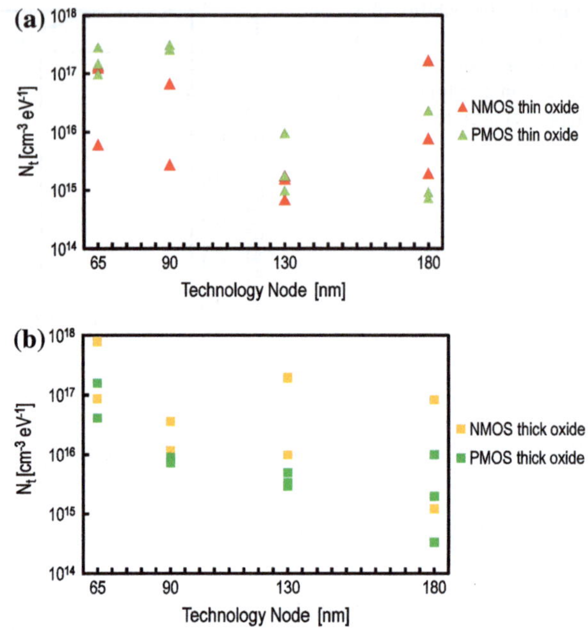

Fig. 9.2 The evolution of the oxide trap density N_t, as a function of the technology node, for thin oxide transistors (**a**) and thick oxide transistors (**b**), based on measurement results reported in design kits from different foundries

to decrease with κ^2 assuming that the oxide trap density N_t remains constant with the technology downscaling. The thermal noise is expected to decrease with κ^2 as well and hence would remain negligible. The ITRS roadmaps expect the oxide trap density to decrease with the technology downscale [10]. Figure 9.2a shows the N_t values, for thin oxide transistors, reported in design kits of three foundries for different technology nodes. It shows that the oxide trap density follows the ITRS roadmap when downscaling from 180 to 130 nm. For more advanced technologies, the oxide trap density stops decreasing but does not show any dramatic increase compared to 180 nm processes. Indeed N_t remains generally in the 10^{16} to 10^{17} cm^{-3}eV^{-1} range.

The measurement results presented earlier [7, 11] explore indirectly the impact of the technology downscaling on the noise reduction. A pixel with a thin oxide SF transistor has been compared to a thick oxide SF based one. For the 180 nm process in which these chips were fabricated, the thick oxide transistor features an oxide capacitance per unit area of 4.5 fF/μm^2 compared to 9.55 fF/μm^2 for the thin oxide transistor. In addition the minimum width set by the design rules is about 0.4 μm for the thick oxide compared to 0.22 μm for the thin oxide transistor. Consequently, using a thin oxide source follower transistor instead of a thick oxide has the same effect than a technology downscale with a scaling factor of 2. Based on this observation the input-referred noise is expected to decrease with a factor of 2 which matches the measurement results shown in Fig. 6.9.

Figure 9.3 shows how the input-referred noise is expected to decrease by only taking advantage of the technology downscale based on (4.28) and the assumption of a constant oxide trap density for deep sub-micron technologies. The starting point

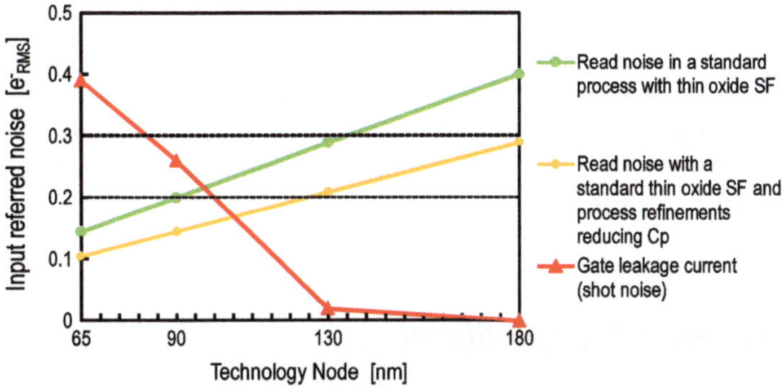

Fig. 9.3 The expected evolution, with technology downscale, of the input-referred $1/f$ noise of CIS designed with standard CMOS process with thin oxide in-pixel SF

corresponds to the input-referred noise measured for the readout chain using a thin oxide SF in a 180 nm CMOS process [7]. It appears that sub-0.3 e^-_{rms} noise can be reached with a 130 nm process and sub-0.2 e^-_{rms} for a 90 nm CIS process. Figure 9.3 also shows the expected input-referred $1/f$ and read noise in case the technology downscaling is combined with process refinements reducing the sense node capacitance C_P. The noise levels for each technology node are obtained using (5.4) and the scaling rules. The starting point corresponds to noise expected when combining the thin oxide PMOS SF with a C_P of 0.25 fF as shown in Fig. 5.10.

9.2.2 *Leakage Current Shot Noise*

Besides the read noise originating from the $1/f$ and thermal noise, the gate leakage current shot noise has been up to now neglected due to the extremely low levels of the leakage currents achieved in the used technology. It is important to investigate the evolution of this noise when using more advanced technologies. Indeed, the gate leakage current increases by several orders of magnitude when downscaling from 180 to 65 nm technologies [9]. Based on (4.45), the shot noise associated to the gate leakage current is hence expected to increase significantly. In order to evaluate its impact, simulations have been performed with transistors having a minimum gate width and length from technologies between 180 nm and 65 nm. The corresponding leakage current shot noise RMS is given by the square root of the total number of electrons crossing the gate in a time interval of 10 μs (enough to read two samples). The results are plotted in Fig. 9.3. Figure 9.3 also shows how the input-referred noise is expected to decrease by only taking advantage of the technology downscale based on (5.4) and below the assumption of constant oxide trap density for deep sub-micron technologies. The starting point corresponds to the input-referred noise obtained using a thin oxide SF in a 180 nm CMOS process [7]. It can be noticed that the

$0.3\,e^-_{rms}$ limit can be crossed if a CIS process is developed with a technology node under 130 nm and a thin oxide transistor is used as a SF. But for technologies under 90 nm, the gate leakage current appears to be a severe problem starting to dominate the total noise. Hence, the optimal technology node is between 130 and 90 nm unless process improvements are applied to reduce the gate leakage current or the CDS sampling period is reduced.

9.2.3 *Random Telegraph Signal*

The reduction of the source follower gate size might increase the probability of RTS noise occurrence [12]. But the amplitude of the RTS noise is inversely proportional to the gate area [13], thus, it would be also reduced by using a minimum source follower gate width. RTS noise may also be a concern with the technology downscaling. In sate-of-the-art low-noise CMOS image sensors, it may result in a dramatically high input referred-noise values of about several e^-_{rms}, but it is only present in a minority of pixels (the tail of the noise histogram). Therefore RTS noise was not accounted for in this work including in the extrapolation towards downscaled technologies because we limited the latter to 65 nm where leakage is much more an issue. A further investigation of input referred noise for such $4T$ pixels in deep submicron technology would definitely require to account for RTS noise. Unfortunately, RTS noise is not modeled in the most common simulators and its occurrence not only depends on the size of the MOS transistors gates but also on the process quality of the foundry since it is directly related to the density of traps in the oxide which are the result of impurities and process defects in that area.

9.3 Summary

The capability of performing photo-electron counting, with an accuracy higher than 90%, using conventional CIS readout chains requires a total read noise level below $0.3\,e^-_{rms}$. This read noise is mainly composed of the $1/f$ noise originating from the in-pixel SF, the thermal noise originating from the pixel and column level saturated transistors, and the shot noise associated to the leakage current at the level of the sense node. The latter is negligible in the technology nodes used currently (above 100 nm).

Based on measurement results reported in recent works, the analytical expressions of the input-referred noise and the downscaling rules, the input-referred read noise for advanced technology nodes is extrapolated. The analysis shows that read noise close to $0.2\,e^-_{rms}$ could be achieved for 90 and 65 nm nodes by implementing standard thin oxide source followers and further reducing the thermal noise with bandwidth reduction or CMS. Further reduction is possible with process refinements reducing the sense node capacitance. Such noise levels would allow accurate photoelectron counting.

The first concern with the technology downscaling are the dramatic increase of the source follower gate leakage current for thin oxide transistors which rises the leakage current shot noise to levels above the $1/f$ and thermal noise for an optimized readout chain. The second concern is the possible increase of the number of pixels featuring RTS noise.

References

1. E. Fossum, The quanta image sensor (qis): concepts and challenges, in *Imaging and Applied Optics*. Optical Society of America (2011), p. JTuE1, http://www.osapublishing.org/abstract.cfm?URI=COSI-2011-JTuE1
2. N. Teranishi, Required conditions for photon-counting image sensors. IEEE Trans. Electron Devices **59**(8), 2199–2205 (2012)
3. D. Bronzi, F. Villa, S. Tisa, A. Tosi, F. Zappa, Spad figures of merit for photon-counting, photon-timing, and imaging applications: a review. IEEE Sensors J. **16**(1), 3–12 (2016)
4. E. Charbon, Single-photon imaging in complementary metal oxide semiconductor processes. Philos. Trans. R. Soc. Lond. A Math. Phys. Eng. Sci. **372**, 2014 (2012)
5. S. Wakashima, F. Kusuhara, R. Kuroda, S. Sugawa, A linear response single exposure cmos image sensor with 0.5e- readout noise and 76ke- full well capacity, in *2015 Symposium on VLSI Circuits (VLSI Circuits)* (2015), pp. C88–C89
6. J. Ma, D. Starkey, A. Rao, K. Odame, E. Fossum, Characterization of quanta image sensor pump-gate jots with deep sub-electron read noise. IEEE J Electron Devices Soc. **3**(6), 472–480 (2015)
7. A. Boukhayma, A. Peizerat, C. Enz, Temporal readout noise analysis and reduction techniques for low-light cmos image sensors. IEEE Trans. Electron Devices **63**(1), 72–78 (2016)
8. M. Seo, S. Kawahito, K. Kagawa, K. Yasutomi, A $0.27e^-_{rms}$ read noise $220\mu v/e^-$ conversion gain reset-gate-less cmos image sensor with $0.11-\mu m$ cis process. IEEE Electron Device Lett. **36**(12), 1344–1347 (2015)
9. L. Lewyn, T. Ytterdal, C. Wulff, K. Martin, Analog circuit design in nanoscale cmos technologies. Proc. IEEE **97**(10), 1687–1714 (2009)
10. Y. Nemirovsky, D. Corcos, I. Brouk, A. Nemirovsky, S. Chaudhry, 1/f noise in advanced cmos transistors. IEEE Instrum. Meas. Mag. **14**(1), 14–22 (2011)
11. A. Boukhayma, A. Peizerat, C. Enz, A 0.4 e-rms temporal readout noise 7.5 m pitch and a 66% fill factor pixel for low light cmos image sensors, in *International Image Sensors Workshop* (2015)
12. P. Martin-Gonthier, P. Magnan, Novel readout circuit architecture for cmos image sensors minimizing rts noise. IEEE Electron Device Lett. **32**(6), 776–778 (2011)
13. H.M. Kwon, I.S. Han, J.D. Bok, S.U. Park, Y.J. Jung, G.W. Lee, Y.S. Chung, J.H. Lee, C.Y. Kang, P. Kirsch, R. Jammy, H.D. Lee, Characterization of random telegraph signal noise of high-performance p-mosfets with a high- k dielectric/metal gate. IEEE Electron Device Lett. **32**(5), 686–688 (2011)

Chapter 10
An Ultra Low Noise CMOS THz Imager

10.1 Introduction

The demand for terahertz (THz) detectors with the capability of delivering real-time imaging increases for an ever broader range of applications such as security [1], non destructive testing [2], medical imaging [3], pharmaceutical applications [4], soil inspection [5] and food inspection [6]. THz electronic detectors are commonly classified into coherent and incoherent detectors. Coherent detectors, also referred to as indirect or heterodyne, are sensitive to the phase and amplitude of the THz radiation. These detectors use a nonlinear device as a mixer to down-convert the THz radiation signal to a lower frequency by means of a local oscillator and sub-harmonic circuit operation [7]. Incoherent detectors, also referred to as direct detectors, are only sensitive to the intensity of the THz radiation but not the phase and hence do not require any local oscillator. Incoherent detectors give less information about the THz radiation but are better suited for building focal plane arrays. Among the THz incoherent detectors such as bolometers [8] or Schottky barrier diodes [9], detectors based on field effect transistors (FET) [10–12] emerge as a key choice for cost-efficiency, low power and on-chip integration.

THz imaging using CMOS FET-based detectors has been mainly performed using a single detector and mechanical scanning of the object [12, 13]. External lock-in amplifiers have been used with modulated THz sources in order to enhance the sensitivity. The first THz CMOS camera featuring a focal plane array of 1 kpixels and operating at video frame rates has been presented in [11]. The purpose of this work is to present the first THz camera including narrow band filtering to enhance the sensitivity without having to use external lock-in amplifiers.

This chapter explores a way to significantly increase the sensitivity of CMOS FET-based THz focal plane arrays by reducing the noise generated by the FET detector and the readout chain. The possibility of easily modulating the THz source above the flicker noise corner frequency of the FET detector and the large pixel pitch, compliant with the THz radiation wavelengths, are exploited by in-pixel integration of highly selective filtering. The band-pass filter required for this application must

A. Boukhayma, *Ultra Low Noise CMOS Image Sensors*,
Springer Theses, https://doi.org/10.1007/978-3-319-68774-2_10

achieve a high Q factor for maximum noise reduction, a tunable central frequency and an easy integration with a low layout footprint. Among the CMOS integrated filters, switched-capacitor (SC) N-path filters seem to be the best candidate to meet these conditions [14].

A 31 × 31 pixels CMOS THz imager operating up to 100 fps is presented. In addition to the antenna and the FET detector, each pixel of the focal plane array integrates an adjustable gain amplification stage in order to adapt the imager to the different THz sources and a tunable high-Q filter made of a combination between a passive SC N-path filter and a broad-band continuous-time (CT) G_m-C filter. The proposed N-path filter is exclusively designed with passive SC network and reaches a Q factor of 100 set by capacitors ratio. This chapter presents a detailed analysis of the readout circuit and particularly the narrow band filter. The analytical results are compared to simulation and measurement results complementing those already presented in [15].

This chapter is organized as follows: Sect. 10.2 reviews the theory of THz detection with MOSFETs. Section 10.3 presents the noise reduction mechanism implemented in this work. In Sect. 10.4, the overall architecture is introduced and the design of the building blocks are described. Section 10.5 presents an analytical analysis of the proposed filter, it shows that the Q factor is simply given by capacitors ratio and describes how the N-path filter is optimized with a broad-band G_m-C filter. Section 10.6 presents the layout implementation of the proposed circuit. In Sect. 10.7 baseband measurement results to characterize the readout chain are presented together with THz test and measurements demonstrating the operation of the proposed imaging technique.

10.2 Operation of CMOS THz Imagers

The mechanism of THz detection using MOSFETs is described by two basic theories. The plasma wave theory and the distributed resistive self mixing. In the early nineties, Dyakonov and Shur described the channel in an idealistic ballistic FET as an electron gas that exhibits a hydrodynamic behavior similar to shallow water [16, 17]. The partial differential equation (PDE) derived from the combination of Euler equation of hydrodynamic movement, the continuity equation and the dependence of the carrier sheet density on the local gate voltage, led them to the description of the propagation of the plasma wave, excited by a THz radiation, in the channel. Classic resistive self mixing is well known in RF applications for excitations below the cutoff frequency of the transistor described by a quasi-static (QS) model. It occurs when both the drain and the gate are coupled to the same radiation. The square-law dependence achieved by this self-mixing results in a DC component proportional to the square of the excitation. Under THz excitation, the non-quasi-static behaviour has to be considered. Distributed resistive self mixing extends the classic resistive self mixing theory [18, 19]. It consists in the division of the channel into small RC segments. Each segment is considered as a QS mixer. Plasma wave theory (in the case of non resonant states) and distributed resistive self mixing conduct to equivalent PDEs [10].

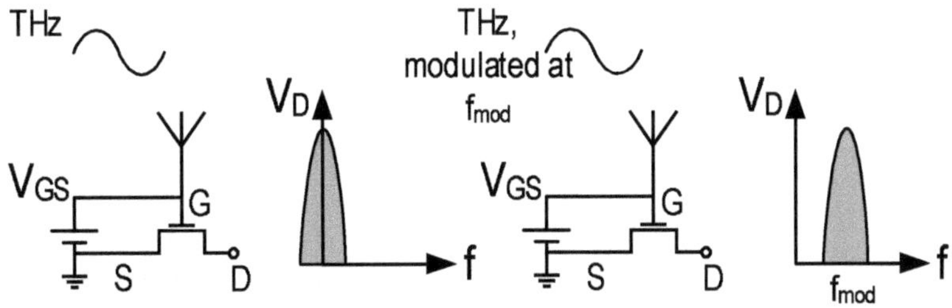

Fig. 10.1 THz rectification with MOSFETs

A practical conclusion of the theoretical approaches and test results is that a FET biased by a constant gate-to-source voltage and excited by a THz signal between its source and gate is expected to exhibit a DC drain to source voltage that depends on the frequency and the amplitude of the THz radiation. When the THz radiation is modulated below the transition frequency of the transistor (about 80 GHz for the 130 nm process used in this work), the DC signal is shifted to the modulation frequency as depicted by Fig. 10.1.

10.3 Noise Reduction Mechanism

The $1/f$ and thermal noise originating from the front-end MOSFET detector limit the signal-to-noise ratio (SNR). The $1/f$ noise can be avoided by shifting the signal above the flicker noise corner frequency and thermal noise can be reduced by limiting the bandwidth. The case of THz imaging offers the possibility of controlling the source that lights the scene. Thus, in order to minimise the $1/f$ noise, the THz source is modulated at a frequency f_{mod} higher than the front-end flicker noise corner frequency. A filter centered at the modulation frequency is then applied to the signal at the output of the MOSFET detector rejecting the $1/f$ noise as depicted by Fig. 10.2. This mechanism acts as chopper stabilization [20]. Based on simulation results of the front-end MOSFET, the flicker noise corner frequency is about 10 kHz. Hence a modulation at 100 kHz is enough to cancel the $1/f$ noise. The thermal noise variance is proportional to the in-pixel filtering bandwidth. Hence, narrow band filtering is highly required. But the filter center frequency must match with the modulation frequency. In order to meet this condition, the modulation phase Φ_{mod} is generated on chip in order to synchronise the filter with the modulation.

The modulation of the THz source can be performed mechanically using choppers or electrically using on-off modulation. It has been shown experimentally that the modulation increases the sensitivity of MOSFET based THz detectors even without implementing any band pass filter after the front-end MOSFET [11, 13]. Moreover, high integration times of a few milliseconds can be reached while achieving high frame rates even with focal plane arrays thousands times larger than the state-of-the-

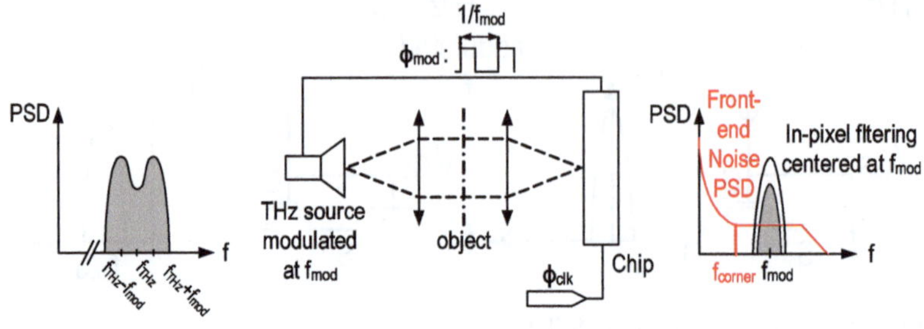

Fig. 10.2 Noise reduction in CMOS THz imagers using source modulation and on-chip filtering

art. Thus very high selective filtering can be implemented without being limited by the filter's rise time. The in-pixel readout chain design based on this noise reduction mechanism is detailed in the following Section.

10.4 Architecture of the Low Noise CMOS THz Imager

10.4.1 Overall Architecture

The presented THz camera is an array of 31 × 31 pixels. The overall block diagram of the imager as well as the pixel are shown in Fig. 10.3. Row and column shift registers are used to address simultaneously the pixels of the array. Each pixel comprises an antenna coupled to a MOSFET detector, a low noise adjustable-gain amplifier and a high-Q filter. As discussed in Sect. 10.3, the pixel includes a digital block that controls the filter and generates the modulation signal Φ_{mod}. In this way, the modulation frequency of the THz source is locked with the center frequency of the filter.

10.4.2 In-Pixel Antenna and Rectifier

Bow tie antennas have been chosen in this design for their broad band characteristics. In addition, these antennas are relatively easy to integrate in the back-end metal of standard CMOS process compared for instance to slot antennas. Log periodic and spiral periodic antennas known for their ultra-broadband characteristics are not easy to integrate under the restrictions and layout design rules. Figure 10.4 shows a layout view of the in-pixel antenna coupled to the NMOS used as a detector. The dimensions of the bow tie antenna have been chosen based on the measurements performed previously in [12, 21] in order to optimize the gain and the impedance matching with the NMOS detector. The radius of the antenna is 64 μm with an angle of $\frac{2}{3}\pi$.

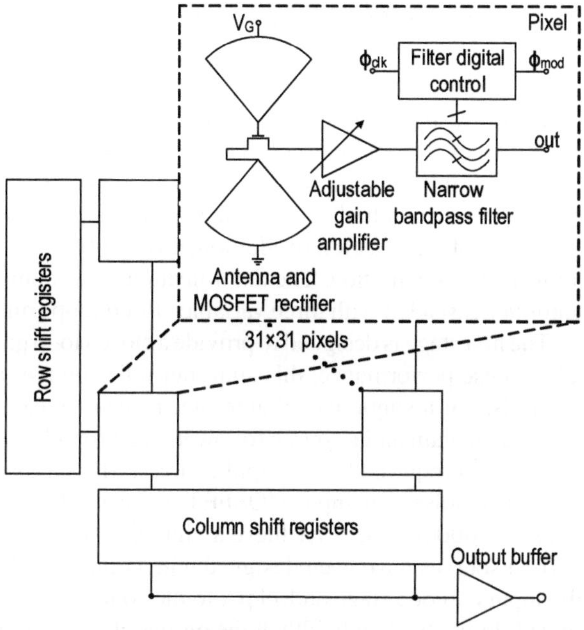

Fig. 10.3 Overall block diagram of the CMOS THz imager

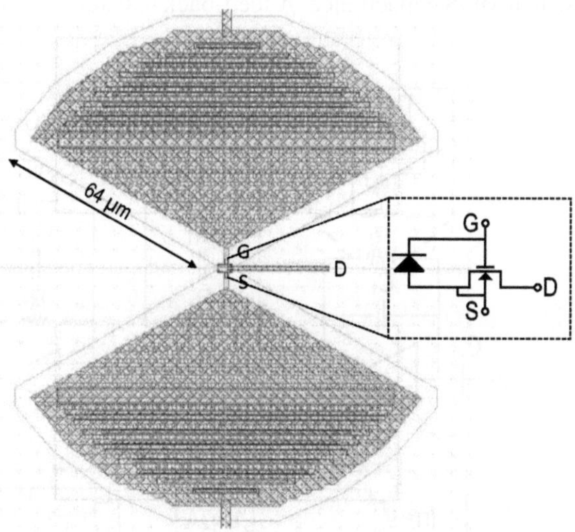

Fig. 10.4 Layout view of the in-pixel THz antenna and MOSFET detector

The two terminals of the antenna are connected respectively to the gate and source of an NMOS transistor used as a FET detector. The two terminals of the antenna are also used to bias the NMOS by applying a gate-to-source voltage. A protection PN junction diode is connected to the antenna in parallel to the NMOS.

10.4.3 In-Pixel Amplification

The baseband voltage generated at the drain node of the MOSFET detector is weak. In fact the responsivity is of the order of a few hundreds of V/W. Standard electrical THz sources deliver Gaussian THz beams of a few mW in the focal plane, thus, at each pixel, the signal at the output of the detector is, at best, expected to be in the order of a few hundreds of µV. Such a weak signal requires a low noise amplifier at the output of the detector. In addition, this amplifier reduces the load of the detector to allow faster modulation. The amplification stage comprises a low noise closed-loop amplifier cascaded with three variable closed loop gain stages as shown by Fig. 10.5.

The first stage is designed to provide a closed loop gain of 32 dB. In order to achieve a low noise performance, this amplifier relies on a minimum number of transistors. It consists of a single-ended amplifier made of a cascode common source stage to provide the gain and a source follower stage for a low output impedance. Since high speed is not required for this application, PMOS transistors have been used for their lower $1/f$ noise. The input MOSFET of the amplifier has been designed with a large W/L of 1600 in order to achieve a high open-loop gain with a minimum noise. The current sources have been designed with compact MOS devices delivering 70 µA to the input cascode stage and 11 µA to the source follower stage. The capacitor C_{cutoff} of 4 pF limits the bandwidth at the output of the common source stage. It is realized with a MOS capacitance. A feed-back resistor of several $G\Omega$ has been used to supply

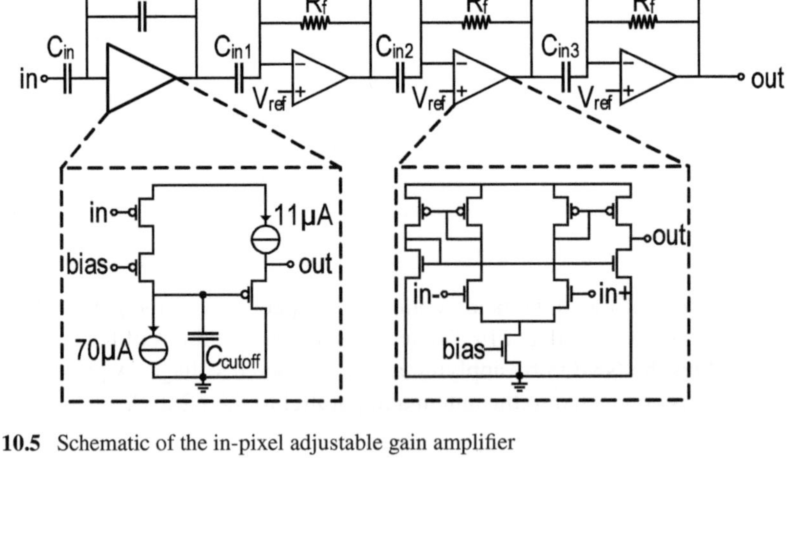

Fig. 10.5 Schematic of the in-pixel adjustable gain amplifier

the gate bias of the common source stage. As shown in Fig. 10.5, it has been realized with two back-to-back PMOS diodes.

The next amplifying stages consist in three OTAs delivering, consecutively, a closed-loop gain of 2, 5 and 10. The gain of each stage can be set to unity using a switched capacitor as shown in Fig. 10.5. The cascaded amplifiers provide a gain that can be varied over a large range in order to adapt the imager to different THz sources. Each of the three stages of the variable gain amplifier feature a simulated phase margin of 60°, a gain margin of 40 dB, a gain bandwidth of 10 MHz and a current consumption of 15 μA.

Based on AC noise simulations of the amplification stage, the input-referred noise PSD of the detector and amplifier is about $5\,\text{nV}/\sqrt{\text{Hz}}$ at 100 kHz and the $1/f$ noise corner frequency is about 10 kHz.

10.4.4 In-Pixel Filtering

The aim of this stage is to select the spectral component centered at the modulation frequency of the THz source. A high-Q filter is needed to achieve high noise rejection as discussed previously. In addition, the filter has to be designed to feature a small layout footprint and a low mismatch in order to achieve low pixel-to-pixel nonuniformity. It is difficult to match those conditions using conventional analog techniques. Active SC circuits offer a good alternative, however they rely on high gain operational amplifiers. In addition to significant power consumption, such active circuits can suffer from stability and sensitivity problems. Consequently, the high-Q filtering is implemented using an N-path passive filter. N-path filters allow the design of band-pass and band-reject filters without the use of inductors. The bandwidth of the N-path filter is independent of the center frequency which is only determined by a clock frequency which makes it easily tunable and robust against mismatch and process variations. The N-path filter also offers a fundamental advantage, the THz modulation frequency and the central frequency of the filter can be generated by the same digital block which make them synchronized. This condition is not easy to meet with conventional CT filters.

In this design, a passive N-path filter, exclusively designed with switched capacitors, has been combined with a CT G_m-C filter. Figure 10.6 shows the schematic of the N-path and G_m-C filters. The SC network of the N-path filter is made of an input capacitor C_R of 40 fF and a network of 16 identical capacitors $C_{P1}...C_{P16}$ of 1.3 pF each. The input capacitor C_R replaces the input resistance in a classical passive SC N-path filter. It samples the signal with a frequency of $16f_{mod}$. The 16 capacitors C_P are cycled with the central frequency of the filter. A digital block composed of frequency dividers generates the different non-overlapping phases controlling the switches out of an external clock signal running at $16f_{mod}$. The corner frequency of the noise at the output of the amplification stage is about 10 kHz, hence, modulation frequencies around few hundreds of kHz are enough to filter out the $1/f$ noise. Consequently, the N-path filter is operated with low frequencies which makes

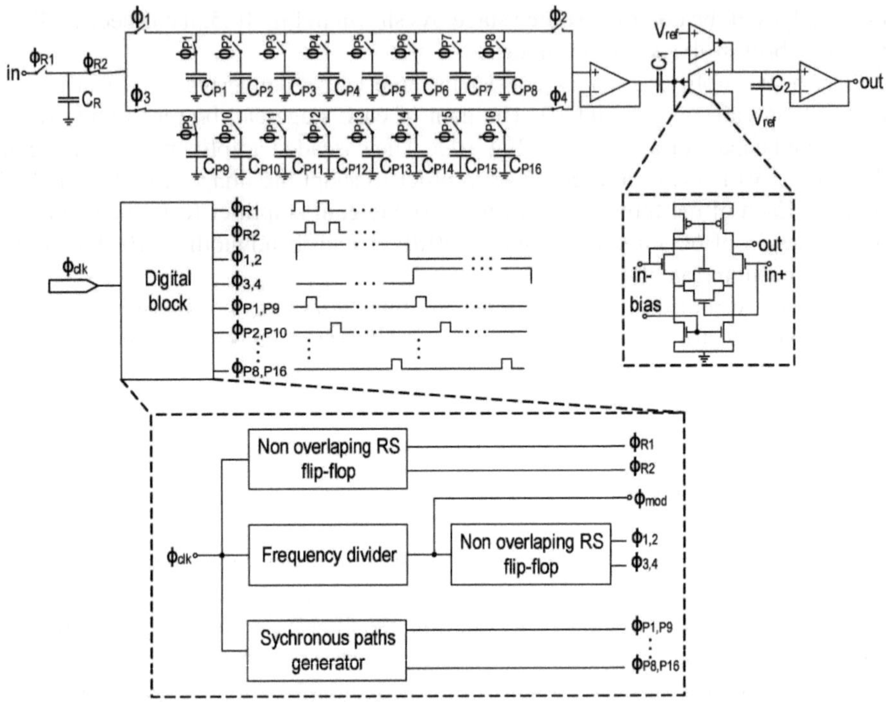

Fig. 10.6 Schematic of the in-pixel high-Q filter

the design of the digital control block and the SC network free of the constraints related to the high frequency operation like the switches non-ideality and the phase noise. As will be discussed in the next Section, a CT G_m-C filter is cascaded with the N-path filter for noise optimization. The transconductors are implemented using degenerated operational transconductance amplifiers (OTAs) for good linearity [22].

10.5 Analysis of the High-Q Filter

10.5.1 Passive SC N-Path Filter

Exhaustive analysis of N-path filters based on switched passive RC networks have been presented in [14, 23, 24]. Here, we aim to derive the transfer function of the passive SC network in an intuitive way and give a simple expression of the transfer function. N-path filters are generally analyzed as linear periodically time-variant systems. The N-path filter design presented in this chapter is a discrete-time (DT) system used to process a CT signal. Figure 10.7 presents the in-pixel readout chain as a signal processing block diagram. The input signal is bandlimited by the amplification

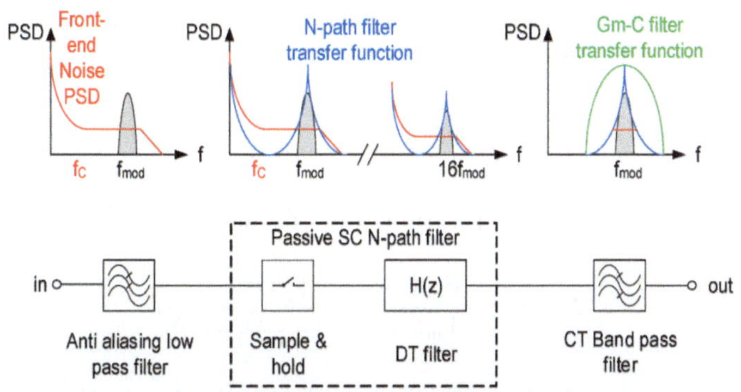

Fig. 10.7 Signal processing block diagram of the readout chain with a spectrum analysis of the filter

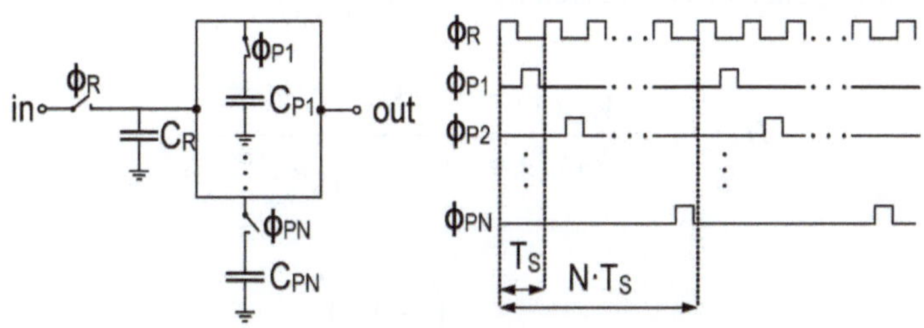

Fig. 10.8 Simplified schematic of the passive SC N-path filter with its timing diagram

stage that acts as an anti-aliasing filter. The CT signal is converted to a DT signal at the input of the N-path filter, processed in DT and then reconstructed at the output using the band pass G_m-C filter.

The DT N-path filter is analyzed using the Z transform. Figure 10.8 shows a simplified schematic of the passive SC N-path filter. We assume that the on resistance of the switches is low enough for a complete settling of the signal during the sampling period T_S. The filter is made of an input capacitor C_R and N identical capacitors C_P. The input CT signal is sampled and held in C_R with a sampling period T_S which is set to be equal to $\frac{1}{N \cdot f_{\text{mod}}}$ in order to match the filter's central frequency with the modulation frequency f_{mod} of the THz source. The capacitors $C_{P1}...C_{PN}$, of the N paths, are cycled with a period of $N \cdot T_S$. At the end of the sampling period $(n-1)T_S$ the input capacitor C_R holds the voltage $V_{\text{in}}((n-1)T_S)$. During the next switching period, one of the capacitors $C_{P1}...C_{PN}$ is connected to the output. It holds the voltage $V_{\text{out}}((n-N)T_s)$ from the previous cycling period $((n-N)T_s)$. Both capacitors are connected to the output node. Thus, they share their charges resulting in an output voltage given by

$$V_{\text{out}}(nT_s) = \frac{C_R V_{\text{in}}((n-1)T_S) + C_P V_{\text{out}}((n-N)T_S)}{C_R + C_P}, \tag{10.1}$$

where $C_P = C_{P1} = ... = C_{PN}$. Consequently, the linear constant-coefficient difference (LCCD) equation of the N-path filter is given by

$$V_{out}(n) = \frac{C_R}{C_R + C_P} V_{in}(n-1) + \frac{C_P}{C_R + C_P} V_{out}(n-N). \tag{10.2}$$

This equation leads to the expression of the Z transform of the transfer function of the filter

$$H_N(z) = \frac{V_{out}(z)}{V_{in}(z)} = \frac{\frac{C_R}{C_R+C_P} z^{-1}}{1 - \frac{C_P}{C_R+C_P} z^{-N}}. \tag{10.3}$$

The equivalent transfer function of the DT N-path filter in Fourier domain is obtained by substituting $e^{-j2\pi \frac{f}{Nf_{mod}}}$ for z since the unity circle is included in the region of convergence of $H(z)$ given by $|z| > \left(\frac{C_P}{C_R+C_P}\right)^{\frac{1}{N}}$.

$$H_N(f) = \frac{V_{out}(f)}{V_{in}(f)} = \frac{\frac{C_R}{C_R+C_P} e^{-j2\pi \frac{f}{Nf_{mod}}}}{1 - \frac{C_P}{C_R+C_P} e^{-j2\pi \frac{f}{f_{mod}}}}. \tag{10.4}$$

Thus,

$$|H_N(f)|^2 = \frac{\left(\frac{C_R}{C_R+C_P}\right)^2}{1 + \left(\frac{C_P}{C_R+C_P}\right)^2 - 2\left(\frac{C_P}{C_R+C_P}\right) cos(2\pi \frac{f}{f_{mod}})}. \tag{10.5}$$

Equation (10.5) suggests that the frequency response of the passive SC filter presented in Fig. 10.8 is a periodic bandpass filter centered at frequencies kf_{mod} for $k < \frac{N}{2}$. The 3 dB bandwidth Δf_{3dB} can be expressed as $(2(f_{3dB} - f_{mod}))$ with $|H(f_{3dB})|^2 = \frac{1}{2}$. The Q factor is then expressed as

$$Q = \frac{f_{mod}}{\Delta f_{3dB}}, \tag{10.6}$$

and the expression of f_{3dB} can be derived as

$$cos\left(2\pi \frac{f_{3dB}}{f_{mod}}\right) = 1 - \frac{1}{2}\left(\frac{C_R}{C_P}\right)^2 \frac{1}{1 + \frac{C_R}{C_P}}. \tag{10.7}$$

In order to simplify (10.7), we consider the case of a high-Q factor for which

$$f_{3dB} - f_{mod} = \frac{1}{2}\Delta f_{3dB} \ll f_{mod}. \tag{10.8}$$

In this case

$$cos\left(2\pi\frac{f_{3\text{dB}}}{f_{\text{mod}}}\right)\simeq 1-\frac{1}{2}\left(\pi\frac{\Delta f_{3\text{dB}}}{f_{\text{mod}}}\right)^2. \tag{10.9}$$

The expression of the Q factor is obtained by combining (10.7) and (10.9) resulting in

$$Q=\pi\frac{C_{\text{P}}}{C_{\text{R}}}\sqrt{1+\frac{C_{\text{R}}}{C_{\text{P}}}}. \tag{10.10}$$

In the case of a high Q factor, $\frac{C_{\text{P}}}{C_{\text{R}}}$ must be higher enough than 1. In this case (10.10) simplifies to

$$Q\simeq\pi\frac{C_{\text{P}}}{C_{\text{R}}}. \tag{10.11}$$

We conclude that the passive SC N-path filter depicted in Fig. 10.8 has a periodic frequency response corresponding to a periodic bandpass filter centered at frequencies kf_{mod} for k lower than $\frac{N}{2}$ and the quality factor of this filter depends only on the ratio $\frac{C_{\text{P}}}{C_{\text{R}}}$. Based on the signal processing block diagram of Fig. 10.7, the power spectral density (PSD) of the signal at the output of the N-path filter can be expressed as

$$S_{\text{out}}(f)=sinc^2(\pi N\cdot f_{\text{mod}})\cdot|H_N(f)|^2\cdot\sum_{n=-\infty}^{+\infty}S_{\text{in}}(f-n\cdot N\cdot f_{\text{mod}}), \tag{10.12}$$

where S_{in} and S_{out} refer respectively to the PSD at the input and output of the SC N-path filter. Note that for the noise PSD, (10.12) only applies for the noise coming from the previous stages and does not include the noise generated by the filter itself. In order to validate the derivation of the transfer function at the output of the N-path filter, the passive SC network shown in Fig. 10.8 has been simulated using Spectre RF©. Figure 10.9 shows the simulated transfer function magnitude compared to the calculation resulting from (10.12) for $N=16$ and $f_{\text{mod}}=125$ kHz. Figure 10.9a shows the frequency response for $\frac{C_{\text{P}}}{C_{\text{R}}}=100$ in the log frequency scale and demonstrates the good matching between the simulation results and (10.12). Figure 10.9b shows the frequency response of the passive SC N-path filter for $\frac{C_{\text{P}}}{C_{\text{R}}}$ set to 1, 10 and 100. The latter figure shows the frequency response in a linear scale with a zoom into the central frequency of the filter which validates (10.10) suggesting that the Q factor only depends on the ratio $\frac{C_{\text{P}}}{C_{\text{R}}}$.

10.5.2 Optimization with a G_m-C Filter

As mentioned in the previous section, the passive SC N-path filter has a periodic frequency response. The baseband spectral component and low frequency noise are therefore not rejected as depicted in Fig. 10.7. Thus for an optimized filtering, an

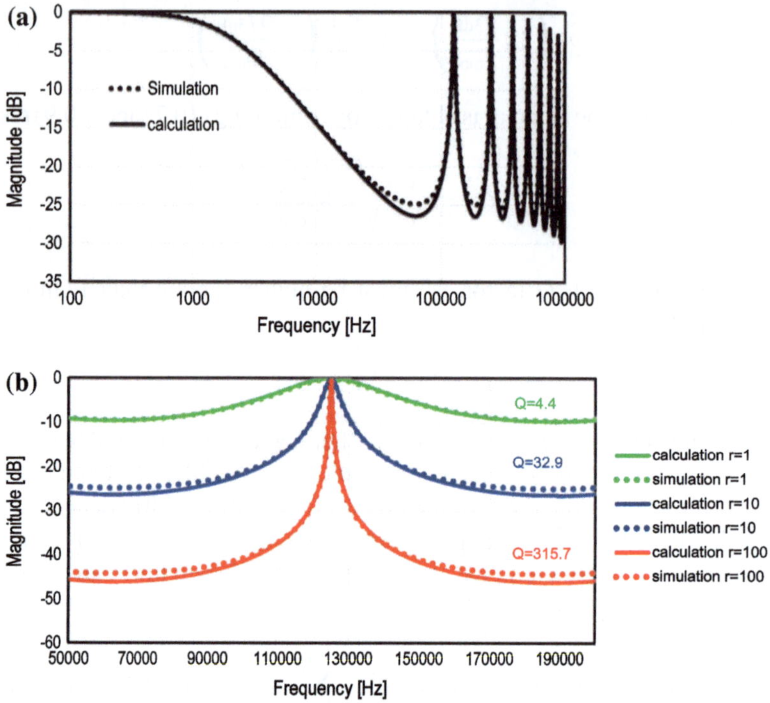

Fig. 10.9 Simulated transfer function of the N-path filter from Fig. 10.8 with Spectre RF© compared to the calculated transfer function from (10.12) for $N = 16$ and $f_{\mathrm{mod}} = 125\,kHz$. **a** shows both transfer functions, in the Log frequency scale with $r = \frac{C_P}{C_R} = 100$, and the impact of the sinc lobes. **b** shows a zoom in a linear scale for $r = \frac{C_P}{C_R}$ set to 1, 10 and 100

additional wide band CT bandpass filter with a lower Q factor has to be applied to the output signal in order to filter out the out-of-band spurious signals and noise. As illustrated in Fig. 10.7, the second filter selects the spectral component centered at f_{mod} and filters out the $1/f$ noise. A Q factor of 1 is enough for the second CT G_m-C filter which is implemented as shown in Fig. 10.6. The square magnitude of the frequency response of this filter is given by

$$|H_{\mathrm{G_m-C}}(f)|^2 = \frac{1}{1 + Q^2\left(\frac{f}{f_0} - \frac{f_0}{f}\right)^2}, \tag{10.13}$$

where the quality factor Q is given by

$$Q = \sqrt{\frac{g_{m1}}{C_1} \cdot \frac{C_2}{g_{m2}}}. \tag{10.14}$$

In this design, $g_{m1} = g_{m2} = g_m$ and $C_1 = C_2 = C$. Thus $Q = 1$. The center frequency f_0 of the G_m-C filter is given by

$$f_0 = \frac{1}{2\pi}\sqrt{\frac{g_{m1} \cdot g_{m2}}{C_1 \cdot C_2}} = \frac{1}{2\pi}\frac{g_m}{C}. \tag{10.15}$$

The latter is set to be equal to the modulation frequency f_{mod}. It is tuned using the bias voltage of the transconductors that determines the value of g_m [22].

10.6 Circuit Implementation

The test chip is fabricated in a standard 130 nm P5M CMOS process. The 31 × 31 pixels array with the row and column circuitry occupy an area of 8 mm×8 mm with a pixel pitch of 240 µm. Both analog and digital parts of the readout chain are powered with a 1.2 V supply. Figure 10.10 shows the chip micrograph with a zoom onto the layout of the pixel. Note that 40% of the pixel area is dedicated to the bow tie antenna integrated using four metal layers. The rest contains the adjustable gain amplifier, the N-path and G_m-C filters, a buffer stage as well as the digital block generating the different phases controlling the switches of the SC N-path filter. Among the readout chain, the N-path filter has the largest layout footprint due to its 16 capacitors of 1.3 pF each integrated using MOS capacitors.

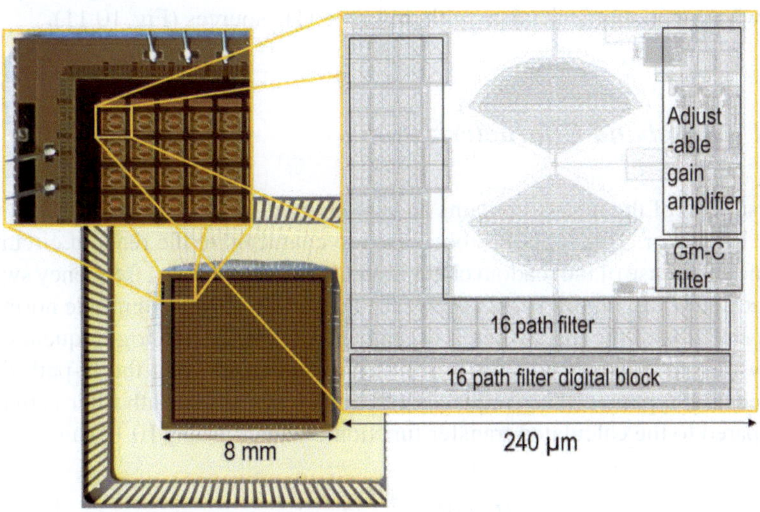

Fig. 10.10 Chip micrograph with a zoom on the pixel layout

Fig. 10.11 Image of the tested chip with PCB test board

10.7 Test and Characterization

The chip test follows two steps: first, a baseband characterization of the readout circuit is performed in order to measure the voltage gain and validate the theoretical analysis and simulation results of the filter. The readout chain total RMS noise voltage $V_{n,RMS}$ is then measured. The second step consists in demonstrating the operation of the whole readout mechanism with different THz sources (Fig. 10.11).

10.7.1 Baseband Characterization

The first pixel of the matrix contains an analog input directly after the THz antenna and FET detector. This input has been used to characterize the readout circuit. The frequency response of the readout chain is obtained by applying a frequency sweep of an input sine wave. Figure 10.12a shows the transfer function magnitude normalized to the peak gain of the filter when it is centered at 156 kHz in a log frequency scale. It shows the measured frequency responses of the filter when the N-path filter is activated and bypassed. The frequency response when the N-path filter is bypassed is compared to the calculated transfer function obtained from (10.13) and given by

$$|H(f)|^2 = \frac{|H_{\mathrm{G_m\text{-}C}}(f)|^2}{1 + (\frac{f}{f_c})^2}, \tag{10.16}$$

where f_c is the cutoff frequency of the output buffer. Figure 10.12b, c show, in a linear frequency scale, the frequency response normalized to the peak gain when

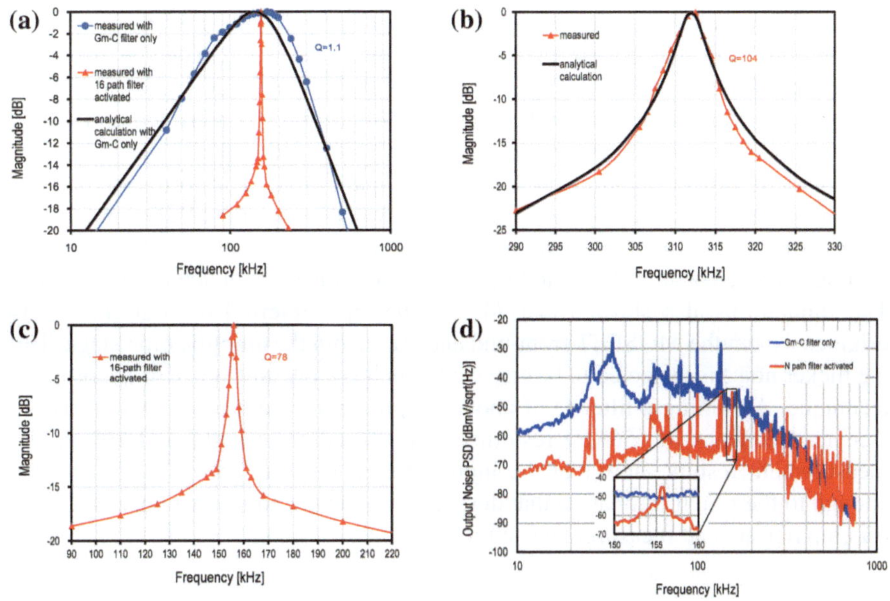

Fig. 10.12 Characterization of the in-pixel baseband circuit with a measured readout chain gain of 58 dB. **a** The transfer function (log frequency scale) when the N-path filter is bypassed, compared to the analytical calculation, and with the N-path filter activated. **b** The measured transfer function (linear frequency scale) when both filters are activated and centered at 312 kHz compared to the analytical calculation. **c** Measured transfer function (linear frequency scale) when both filters are activated and centered at 156 kHz. **d** The output noise PSD obtained when both filters are activated and when the N-path filter is bypassed with a zoom onto the center frequency of the N-path filter

the N-path filter is activated and centered at 312 kHz and 156 kHz respectively. The frequency response when the N-path filter is centered at 312 kHz is compared to the calculated transfer function obtained from (10.12) and (10.13) given by

$$|H(f)|^2 = \frac{sinc^2(\pi N \cdot f_{\mathrm{mod}}) \cdot |H_N(f)|^2 \cdot |H_{\mathrm{G_m-C}}(f)|^2}{1 + (\frac{f}{f_c})^2}, \tag{10.17}$$

for $N = 16$ and $\frac{C_P}{C_R} = 100/\pi$. Figure 10.12a and c demonstrate the extremely high selectivity of the filter having a Q factor of 100. It also shows the excellent match between the theoretical model presented in Sect. 10.5 and the measurement results. Note that a loss of 8% has been measured when the N-path filter is activated due to the passive implementation of the SC N-path filter.

Figure 10.12d shows the broadband noise power spectral density (PSD) measured with a spectrum analyzer at the output of the readout chain when the N-path filter is activated and bypassed. The broadband noise PSD is clearly shaped by the frequency response of the G_m-C filter when the N-path filter is bypassed. For this measurement, the N-path filter has been centered at 156 kHz. Figure 10.12d shows that when the

N-path filter is activated, the noise PSD is reduced by 20 dB out of the narrow SC N-path filter band. The output RMS noise voltage has been calculated by integrating the measured PSDs over the frequency range [0:500 kHz] (the bandwidth of the output buffer) resulting in 0.2 mV when the N-path filter is activated and 1.9 mV when bypassed. These measurement have been performed with a measured readout chain gain of 58 dB (including the loss of the N-path filter). The corresponding total input-referred noise is as low as 0.2 μV. The simulation results of the amplification stage show that the input referred noise PSD of the detector and amplification stages is about 5 nV/$\sqrt{\text{Hz}}$ at 100 kHz and the $1/f$ noise corner frequency is about 10 kHz. The simulations also showed that 74% of the input-referred readout chain noise originates from the MOSFET front-end and 20% from the amplification stage. Thus the flicker noise is expected to be canceled which is a big advantage of the proposed technique. Assuming a filter Q of about 100, the bandwidth is about 1.56 kHz for a center frequency of 156 kHz resulting in a thermal noise voltage of 197 nV. The measurement results show that the input referred noise is 248 nV which is close enough to the rough estimation that does not include the additional noise due to the off chip buffers and measurement setup.

10.7.2 THz Characterization

A characterization of the presented THz imager has been performed with different active sources in order to validate the readout scheme depicted in Fig. 10.2 and measure the responsivity of the imager with a modulated THz radiation. The imager responsivity is measured as the ratio between the total output voltage of all the pixels of the imager and the power of the THz radiation incident to the imager position. It is expressed in V/W as

$$R_v = \frac{\Sigma_{i=1}^{961} V_{\text{out},i}}{P_{\text{THz}}}, \tag{10.18}$$

where $V_{\text{out},i}$ refers to the voltage amplitude at the output of the ith pixel of the imager and P_{THz} refers to the measured THz power incident to the chip area. Note that this measurement method gives an absolute responsivity taking into account the fill factor and the antenna directivity.

The sensitivity of the imager is then expressed as a total noise equivalent power NEP_{total} in Watts

$$NEP_{\text{total}} = \frac{V_{n,RMS}}{R_v}. \tag{10.19}$$

The NEP_{total} represents the amount of THz power corresponding to an SNR of 1 without having to take into account any other off-chip considerations. Note that NEP_{total} differs from the NEP. The NEP which is given in pW/$\sqrt{\text{Hz}}$ describes better the noise performance of pixels only comprising a detector and low-noise amplifier and for which the bandwidth control is performed off-chip by means of a

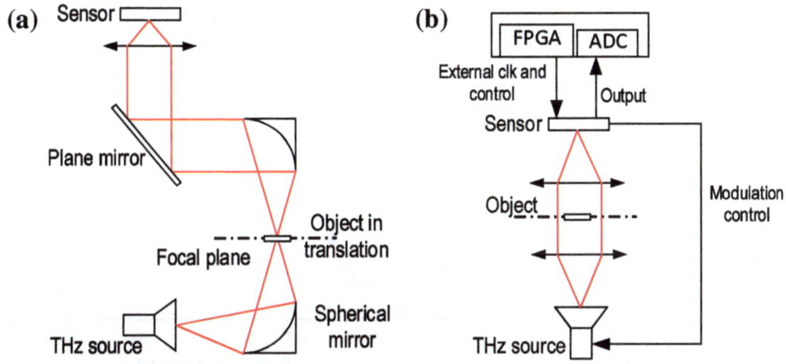

Fig. 10.13 Experimental setup used with **a** the 2.5 THz laser gas source and **b** with the multiplying chain sources in the range [200:600] GHz

lock-in amplifier for instance [13]. In this work the selective filtering is performed in-pixel. The global performance in this case is given in NEP_{total} as reported in [11]. The NEP value can be obtained from the noise spectral density at the modulation frequency which is also the center frequency of the filter.

Two continuous-wave (CW) frequency multiplying chains are used to generate the THz radiation between 200 and 600 GHz. The experimental setup used with these CW sources is depicted in Fig. 10.13. The THz sources are modulated electrically (On/Off modulation) using the modulation phase Φ_{mod} generated by the on-chip digital block in order to be synchronous with the center frequency of the in-pixel high selective filtering. Two lenses sharing the same optical axis are positioned between the THz source and the tested imager. A distance close to the lenses focal length separates the CW THz source and the imager from the corresponding lenses in order to make sure that most of the THz power generated by the CW source is projected on the tested focal plane array. A powermeter based on a Schottky barrier diode is then used to measure the total available power at the chip position. The CW 200 GHz source delivers a power of 2 mW at the chip position and the one used to generate THz radiations in the range of [270–600 GHz] delivers a power of 1 mW at 270 GHz and 0.5 mW at 600 GHz.

A gas laser 2.5 THz source has also been used in order to characterize the THz imager at the higher edge of the THz band. Unfortunately, electrical modulation is not possible with such a source, and the mechanical chopping operates at frequencies below hundreds of Hertz. Thus the characterization method described previously could't be applied in this case.

First, the test pixel was used in order to identify the gate bias voltage corresponding to the maximum responsivity. Figure 10.14a shows the voltage at the output of the MOSFET detector (between the drain and source), normalized to its maximum value, as a function of the gate bias voltage (gate-to-source). For THz radiations of 200 GHz and 2.5 THz, the amplitude of the output voltage reaches its maximum for a gate bias

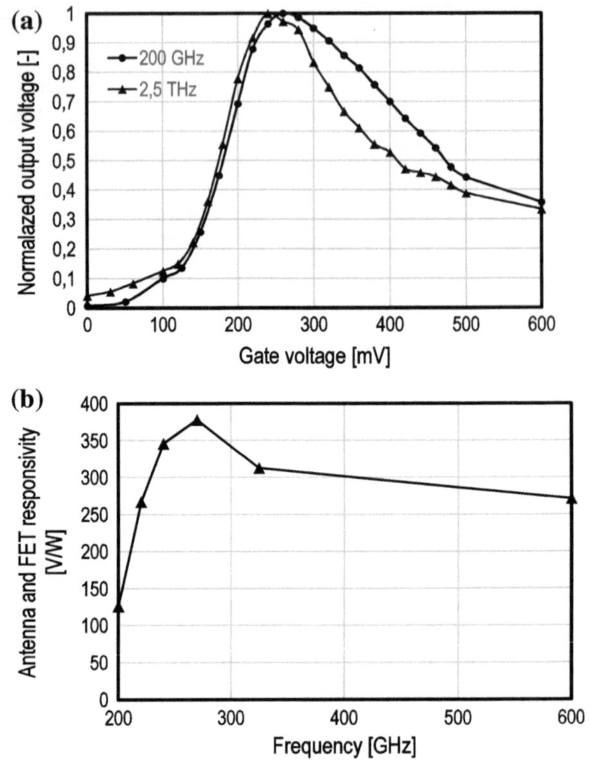

Fig. 10.14 THz characterization of the sensor: **a** The amplitude of the output voltage, normalized to its maximum, over the gate bias for a THz radiations at 200 GHz and 2.5 THz. **b** The measured responsivities for frequencies between 200 and 600 GHz

of 0.25 V. This voltage is close to the threshold voltage of the MOSFET used in this design.

Figure 10.14b shows the responsivities obtained using (10.18) with the CW multiplying chain THz sources modulated at 156 kHz with the imager readout chain set to 58 dB. It wasn't possible to obtain measurements in the full 3 dB bandwidth of the antenna, but this measurement shows that it is larger than 400 GHz and confirms the broadband characteristic of the bow tie antenna. These measurements were used to calculate NEP_{total} based on (10.19). The NEP_{total} values are 0.6 nW at 270 GHz and 0.8 nW at 600 GHz.

In order to obtain the NEP value in pW/$\sqrt{\text{Hz}}$, the measured noise PSD at 156 kHz (shown in the inset of Fig. 10.12d) is used to obtain the output noise in V/$\sqrt{\text{Hz}}$, this value is then divided by R_v in order to calculate the NEP. From the inset of Fig. 10.12d, showing a zoom onto the center frequency of the filter, the measured output noise at the center frequency is −45 dBmV/$\sqrt{\text{Hz}}$ corresponding to 5.6 µV/$\sqrt{\text{Hz}}$. The resulting NEP value is then 18.7 pW/$\sqrt{\text{Hz}}$ at 270 GHz. Similarly a NEP of 25.9 pW/$\sqrt{\text{Hz}}$ is obtained at 600 GHz accounting for the same PSD value and a slightly reduced R_v of 216 kV/W.

Figure 10.15a shows the image of the 200 GHz Gaussian beam. It can be noticed that the beam is not large enough to cover the whole imager. Thus, making good

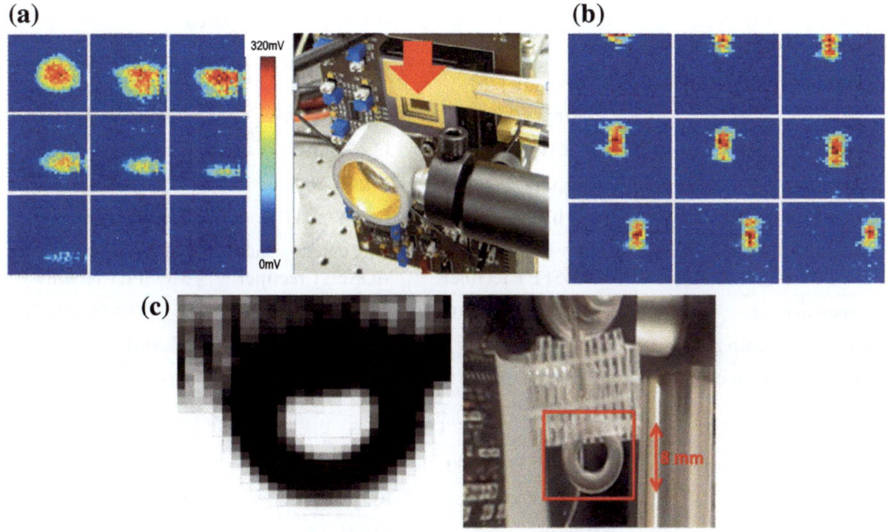

Fig. 10.15 **a** Video sequence, obtained at 100 fps, of a copper ruler passing in front of a 200 GHz beam of 2 mW measured at the imager position and modulated at 156 kHz. **b** Video sequence, obtained at 100 fps, when translating vertically then horizontally the imager exposed to a 270 GHz source with a rectangular waveguide. The THz source is modulated at 156 kHz and delivers 0.5 mW at the sensor position. **c** Image of a metallic ring held with tape obtained with the 2.5 THz using the setup of Fig. 10.13, the antenna and FET detector of one pixel and a lock-in amplifier

images without mechanical scanning is still difficult with such levels of THz radiation power. Figure 10.15a also shows images from a video sequence obtained, at a frame rate of 100 fps, when passing a coper ruler in front of the imager to reflect the THz beam. Figure 10.15b shows images from a video sequence, obtained at 100 fps, when translating vertically then horizontally the imager exposed to a 270 GHz source with a rectangular waveguide. The THz source is modulated at 156 kHz and delivers 0.5 mW. The characterization of the imager with the 2.5 THz source was not possible for the reasons mentioned above, thus, only the antenna and FET detector part of one pixel from the array was used in order to obtain an image by performing a mechanical scan. Figure 10.13a shows the experimental setup used for this operation. The object is positioned in the focal point between two spherical mirrors. The THz radiation is then collimated by the second mirror and focused on the sensor. Figure 10.15c shows the image of a metal ring held with tape, using the setup described above, by mechanically scanning the object in the focal plane between the two spherical mirrors. The THz power measured at the sensor position is 10 mW. The readout were performed with a lock-in amplifier and a mechanical chopping of 290 Hz.

The global performance of the presented THz image sensor is summarized in Table 10.1 and compared to recently reported THz focal plane arrays. Note that the power consumption of the presented sensor is higher than other works. This is mainly due to power consumption of the closed loop amplifiers. The power consumption

Table 10.1 Overview of the presented THz imager performance compared to recently reported CMOS THz imagers

Process	This work [25] 130 nm CMOS	Ref. [11] 65 nm CMOS	Ref. [13] 130 nm CMOS
Array	31×31 pixel	32×32 pixel	8× 8pixel
Pixel Pitch	240 μm×240 μm	80 μm×80 μm	83 μm×83 μm
Power Consumption	174 μW/pixel	2.5 μW/pixel	150 μW/pixel
Detection Device	Bow tie antenna and MOSFET rectifier	Ring antenna and MOSFET rectifier	Patch antenna and MOSFET rectifier
Responsivity R_v	300 kV/W@270 GHz	115 kV/W@860 GHz	3.4 kV/W@820 GHz
including on-chip gain of	58 dB	50 dB	13.5 dB
Pixel and readout chain design	Closed loop adjustable gain	Open loop gain	Open loop gain
	High selective filtering $Q = 100$		
NEP	$18.7\,\text{pW}/\sqrt{\text{Hz}}$ [a]	$100\,\text{pW}/\sqrt{\text{Hz}}$	$15.5\,\text{pW}/\sqrt{\text{Hz}}$
NEP_{total}	0.6 nW	12 nW	

[a] The NEP measured at the center frequency of the filter

of the digital block controlling the SC N-path filter can be reduced by sharing the digital control signals between multiple pixels. This work presents a relatively higher pixel pitch due to the in-pixel additional circuitry. The fill factor can be improved in different ways: for example (a) the additional amplification stages can be omitted and (b) the digital block of the N-path filter can be moved outside the pixel by making it shared by a higher number of pixels. A smaller capacitance C_R can be used instead of 40 fF in order to use smaller capacitance C_P for the same Q factor. In addition the more advanced technologies than 130 nm have higher MOS capacitance per unit area. Thus, the layout footprint of the filter can be dramatically reduced without degrading the Q factor taking advantage of CMOS technology down-scaling.

10.8 Summary

This work introduces a noise reduction mechanism for CMOS THz focal plane arrays that consists in modulating the THz radiation at the source level and performing highly selective filtering at the pixel level. The noise reduction is achieved by rejecting the low frequency noise and drastically reducing the bandwidth of the thermal noise. A 31×31 pixels THz imager is fabricated in a 130 nm CMOS standard process has been presented. The imager pixels include a bow tie antenna, a low noise amplifier and a high-Q filter. The latter consists in a passive SC N-path filter followed by a CT G_m-C filter. A simplified analytical calculation of the filter transfer function is provided. The test and characterization of the presented chip validates the theoretical analysis of the filter. A high Q of 100 has been reached. The input-referred readout noise has been measured to be as low as 0.2 μV RMS and a drastic readout noise reduction has been demonstrated as expected theoretically. The noise reduction scheme has been validated by testing the imager with different THz sources in the [200–600] GHz range. A total NEP as low as 0.6 nW at 270 GHz has been measured. The results presented with this test chip can be improved at both sensitivity and power consumption levels. Indeed, the fill factor of the presented pixel can be improved by omitting the additional gain stages as well as the digital blocks. The presented readout chain can be used with a more performant antenna with a better responsivity. The noise reduction of the readout chain relies on the very high Q factor of the filter which is directly related to the area of the MOS capacitors. Thus, the design can also be improved by using a more advanced CMOS process with a higher oxide capacitance density that would allow the same selectivity for a smaller area in order to increase the fill factor.

References

1. H.-B. Liu, H. Zhong, N. Karpowicz, Y. Chen, X.-C. Zhang, Terahertz spectroscopy and imaging for defense and security applications. Proc. IEEE **95**(8), 1514–1527 (2007)
2. E. Yakovlev, K. Zaytsev, I. Dolganova, S. Yurchenko, Non-destructive evaluation of polymer composite materials at the manufacturing stage using terahertz pulsed spectroscopy. IEEE Trans. Terahertz Sci. Technol. **5**(5), 810–816 (2015)
3. Z. Taylor, R. Singh, D. Bennett, P. Tewari, C. Kealey, N. Bajwa, M. Culjat, A. Stojadinovic, H. Lee, J. Hubschman, E. Brown, W. Grundfest, Thz medical imaging: in vivo hydration sensing. IEEE Trans. Terahertz Sci. Technol. **1**(1), 201–219 (2011)
4. J.-Y. Kim, R. Boenawan, Y. Ueno, K. Ajito, Quantitative mapping of pharmaceutical cocrystals within cellulose by terahertz spectroscopy. J. Lightwave Technol. **32**(20), 3768–3773 (2014)
5. V. Dworak, S. Augustin, R. Gebbers, Application of terahertz radiation to soil measurements: initial results. Sensors **11**(10), 9973 (2011), http://www.mdpi.com/1424-8220/11/10/9973
6. S.-T. Han, W.K. Park, H.S. Chun, Development of sub-thz gyrotron for real-time food inspection, in *2011 36th International Conference on Infrared, Millimeter and Terahertz Waves (IRMMW-THz)* (2011), pp. 1–2
7. F. Friederich, W. von Spiegel, M. Bauer, F. Meng, M. Thomson, S. Boppel, A. Lisauskas, B. Hils, V. Krozer, A. Keil, T. Loffler, R. Henneberger, A. Huhn, G. Spickermann, P. Bolivar, H. Roskos, Thz active imaging systems with real-time capabilities. IEEE Trans. Terahertz Sci. Technol. **1**(1), 183–200 (2011)
8. F. Simoens, J. Meilhan, J.-A. Nicolas, Terahertz real-time imaging uncooled arrays based on antenna-coupled bolometers or fet developed at cea-leti. J. Infrared Millimeter Terahertz Waves **36**(10), 961–985 (2015). https://doi.org/10.1007/s10762-015-0197-x
9. R. Han, Y. Zhang, Y. Kim, D.Y. Kim, H. Shichijo, E. Afshari, K. Kenneth, Active terahertz imaging using schottky diodes in cmos: array and 860-ghz pixel. IEEE J. Solid-State Circuits **48**(10), 2296–2308 (2013)
10. W. Knap, V. Kachorovskii, Y. Deng, S. Rumyantsev, J.-Q. L, R. Gaska, M.S. Shur, G. Simin, X. Hu, M.A. Khan, C.A. Saylor, L.C. Brunel, Nonresonant detection of terahertz radiation in field effect transistors. J. Appl. Phys. **91**(11), 9346–9353 (2002). https://doi.org/10.1063/1.1468257
11. R. Al Hadi, H. Sherry, J. Grzyb, Y. Zhao, W. Forster, H. Keller, A. Cathelin, A. Kaiser, U. Pfeiffer, A 1 k-pixel video camera for 0.7-1.1 terahertz imaging applications in 65-nm cmos. IEEE J. Solid-State Circuits **47**(12), 2999–3012 (2012)

12. F. Schuster, H. Videlier, A. Dupret, D. Coquillat, M. Sakowicz, J. Rostaing, M. Tchagaspanian, B. Giffard, W. Knap, A broadband thz imager in a low-cost cmos technology, in *2011 IEEE International Solid-State Circuits Conference Digest of Technical Papers (ISSCC)* (2011), pp. 42–43
13. D.Y. Kim, S. Park, R. Han, K. Kenneth, 820-ghz imaging array using diode-connected nmos transistors in 130-nm cmos, in *2013 Symposium on VLSI Circuits (VLSIC)* (2013), pp. C12–C13
14. A. Ghaffari, E. Klumperink, M. Soer, B. Nauta, Tunable high-q n-path band-pass filters: Modeling and verification. IEEE J. Solid-State Circuits **46**(5), 998–1010 (2011)
15. A. Boukhayma, J.-P. Rostaing, A. Mollard, F. Guellec, M. Benetti, G. Ducournau, J.-F. Lampin, A. Dupret, C. Enz, M. Tchagaspanian, J.-A. Nicolas, A 533pw nep 31 × 31 pixel thz image sensor based on in-pixel demodulation, in *ESSCIRC 2014-40th European Solid State Circuits Conference (ESSCIRC)* (2014), pp. 303–306
16. M. Dyakonov, M. Shur, Detection, mixing, and frequency multiplication of terahertz radiation by two-dimensional electronic fluid. IEEE Trans. Electron Devices **43**(3), 380–387 (1996)
17. M. Dyakonov, M. Shur, Shallow water analogy for a ballistic field effect transistor: New mechanism of plasma wave generation by dc current. Phys. Rev. Lett. **71**, 2465–2468 (1993). https://doi.org/10.1103/PhysRevLett.71.2465
18. A. Lisauskas, U. Pfeiffer, E. jefors, P.H. Bolvar, D. Glaab, H.G. Roskos, Rational design of high-responsivity detectors of terahertz radiation based on distributed self-mixing in silicon field-effect transistors. J. Appl. Phys. **105**(11), 2009. https://doi.org/10.1063/1.3140611
19. E. Ojefors, U. Pfeiffer, A. Lisauskas, H. Roskos, A 0.65 thz focal-plane array in a quarter-micron cmos process technology. IEEE J. Solid-State Circuits **44**(7), 1968–1976 (2009)
20. C. Enz, G. Temes, Circuit techniques for reducing the effects of op-amp imperfections: autozeroing, correlated double sampling, and chopper stabilization. Proc. IEEE **84**(11), 1584–1614 (1996)
21. F. Schuster, D. Coquillat, H. Videlier, M. Sakowicz, F. Teppe, L. Dussopt, B. Giffard, T. Skotnicki, W. Knap, Broadband terahertz imaging with highly sensitive silicon cmos detectors. Opt. Express **19**(8), 7827–7832 (2011), http://www.opticsexpress.org/abstract.cfm?URI=oe-19-8-7827
22. F. Krummenacher, N. Joehl, A 4-mhz cmos continuous-time filter with on-chip automatic tuning. IEEE J. Solid-State Circuits **23**(3), 750–758 (1988)
23. D. von Grunigen, R. Sigg, J. Schmid, G. Moschytz, H. Melchior, An integrated cmos switched-capacitor bandpass filter based on n-path and frequency-sampling principles. IEEE J. Solid-State Circuits **18**(6), 753–761 (1983)
24. M. Darvishi, R. van der Zee, B. Nauta, Design of active n-path filters. IEEE J. Solid-State Circuits **48**(12), 2962–2976 (2013)
25. A. Boukhayma, A. Dupret, J.-P. Rostaing, C. Enz, A low-noise cmos thz imager based on source modulation and an in-pixel high-q passive switched-capacitor n-path filter. Sensors **16**(3), (2016), http://www.mdpi.com/1424-8220/16/3/325

Chapter 11
Conclusion

11.1 Read Noise Reduction in CMOS Image Sensors

The performance of CMOS image sensors under low light conditions is limited by the readout chain noise. The reduction of this noise to deep sub-electron levels in standard processes is crucial for introducing CIS in new applications requiring photoelectron counting or single photoelectron detection capability. With an input-referred readout noise approaching $0.3\,e^-_{rms}$, CIS would be capable of photoelectron counting.

The read noise in CIS results from the superposition of the random fluctuations generated by the electronic devices at different levels of the readout chain. These include thermal noise, $1/f$ and RTS noise and leakage current shot noise. The leakage current shot noise is dominated by the source follower gate tunneling and sense node junction leakage current. This noise is negligible for state-of-the art CIS processes but could rise significantly in advanced processes.

11.1.1 General Methodology for Read Noise Reduction in CIS

Introducing the gain at the earliest stage of the readout chain is the first step to limit the number of dominating noise sources. The gain can be introduced at the pixel-level by operating the in-pixel readout transistor in the common source configuration. The second possibility is to use a source follower stage at the pixel level and implement the gain at the column level. The second option is more practical and robust against process spacial variations and reset switch non-ideality. In addition, the comparison of both configurations from the noise performance perspective shows that the common source configuration has a slight advantage over the source follower configuration in terms of thermal noise.

The implementation of the column-level gain makes the noise contribution of the following stages like analog buffers, sample-and-holds and ADC negligible and limits the noise analysis to the first two blocks.

A. Boukhayma, *Ultra Low Noise CMOS Image Sensors*,
Springer Theses, https://doi.org/10.1007/978-3-319-68774-2_11

11.1.1.1 Thermal Noise Reduction

The thermal noise can be reduced to deep sub-electron noise levels using the following techniques:

- Controlling the bandwidth of the readout chain at the level of the column-level amplifier. For an OTA based switch capacitor amplifier, when the closed loop gain is higher than 10, reducing the OTA transconductance, increasing the closed loop gain, or increasing the sum of the load and feedback capacitors have all equivalent impact on the thermal noise reduction.
- The reduction of the excess noise factor of the column-level OTA. This can be achieved by designing the OTA with a minimum number of noisy transistors. The fully cascoded common source amplifier is well suited for this application. Indeed, the cascode transistors do not contribute significantly to the total noise. In this case the excess noise factor can be further decreased by designing the load transistor of the common source stage to have a lower transconductance than the amplifying transistor.
- Increasing the CMS order M. Note that the M and the bandwidth of the column-level amplifier have both equivalent impacts on the thermal noise reduction: a higher CMS order requires a higher readout time for a given sampling frequency which is equivalent to a bandwidth reduction by the same order.
- Using a minimum sized in-pixel source follower and a minimum sense node capacitance.
- Reducing the thermal noise excess factor of the source follower stage by reducing the transconductance of the load transistor. This can be achieved by designing it to have a lower aspect ratio than the in-pixel source follower transistor.

11.1.1.2 $1/f$ Noise Reduction

By implementing these techniques, the $1/f$ noise of the readout chain becomes the dominant noise source. The first $1/f$ noise reduction comes from the readout scheme of CIS based on double sampling. Even with a CDS, the residual $1/f$ noise still dominates the readout chain noise when the thermal noise is reduced to deep sub-electron levels. The residual $1/f$ noise can be reduced using the following techniques:

- Designing all the transistors of the readout chain located outside the pixel to have a gate area much larger than the in-pixel source follower in order to make their contribution to the total $1/f$ noise negligible. This has a negligible impact on the chip area since all these transistors are common to the columns and the chip area is dominated by the pixels array.
- Designing the in-pixel source follower to have an optimal gate width and length. The optimal values depend on the technological parameters of the foundry and correspond generally to a minimum gate width allowed by the technology and an gate length slightly larger.

- Increasing the oxide capacitance per unit area of the in-pixel source follower transistor.
- Reducing the oxide trap density of the in-pixel source follower transistor.
- Reducing the sense node capacitance. This can be performed by minimizing the sense node junction area, reducing the metal wires coupling capacitance with the sense node and reducing the overlap between the sense node and the transfer and reset gates.
- Performing CMS with an order between 2 and 8 for a lower residual $1/f$ noise than a simple CDS.

11.1.2 The Transistor Oxide Thickness: A Degree of Freedom for Noise Optimization

In this work, a way has been suggested to match all the points related to pixel design regarding the $1/f$ noise reduction in a standard CIS process without any process refinements. It consists in the selection of a thin oxide PMOS source follower instead of the traditionally used thick oxide NMOS transistors. Indeed, the foundries offer at least two options for the oxide thickness: thick oxide transistor for high voltage applications like IOs and thin oxide "digital" transistors. The thin oxide transistors allow a smaller minimum gate width, have higher oxide capacitance, and feature generally lower oxide trap density. Hence, selecting a thin oxide PMOS source follower combines three $1/f$ noise reduction techniques.

11.1.3 Thin Oxide Source Follower Implementation for Sub-0.5-electrons Noise in a Standard CIS Process

- The implementation of a thin oxide source follower in a high voltage (3.3 V) circuit requires a careful design. A successful circuit implementation has been presented allowing the operation of the thin oxide transistor with a voltage difference between all its four nodes not exceeding 1.8 V without sacrificing the pixel voltage swing. Pixels with 7.5 μm pitch and 66% fill factor using the proposed in-pixel thin oxide source followers have been implemented successfully on the same test chip, fabricated in a 180 nm CIS process, together with pixels with thick oxide buried channel source follower already optimized at the process level for low noise by the foundry. Both pixels share the same column level amplifiers and CDS. The column-level amplifier has been designed using the noise reduction techniques mentioned above and limits the bandwidth optimally for a CDS of 5 μs. The measurements have been performed successfully. The newly proposed pixels feature a mean noise of 0.4 e^-_{rms} with a spatial standard deviation over the pixels of 0.08 e^-_{rms} while the low noise pixels proposed by the foundry feature a mean noise of 0.9 e^-_{rms} and a standard deviation of 0.24 e^-_{rms}. These measurements have been obtained on a population of 85 pixels of each type. The measurements also show that the implementation of the thin oxide transistor does not degrade significantly the pixel dynamic range and allows a voltage swing above 1 V at the pixel output.

- A second 25 mm^2 chip embedding a full VGA imager implementing in-pixel thin oxide PMOS source followers has been fabricated in the same 180 nm 4PM CIS process. The proposed imager features dual column-level gain and column parallel SSAD reaching 80 fps. For this chip, in addition to the optimal design for low noise, a compact layout with a common n-well for the PMOS transistors of neighboring pixels and optimized for minimum sense node parasitic capacitance has been presented. The pixel pitch is 6.5 μm with a fill factor of 40%.
  The test and characterization of the VGA imager have demonstrated that sub-0.5-electron noise, in a full VGA APS, can be achieved using a standard CIS process by a proper circuit noise optimization exploiting all the degrees of freedom left to the designer for minimizing the total input-referred noise. The measurements resulted in the majority of pixels peaking at 0.48 e^-_{rms} and a minority of pixels showing an RTS noise of a few electrons. The input-referred TRN has been measured without CMS and at a short pixel (line) readout time, for a sub-electron read noise CIS, of 25 μs.
  This work has also provided a full characterization of the VGA imager including a PTC curve pointing out a conversion gain of 160 μV/e^-, a full well capacity of 6400 e^- in dual gain mode, a photo response non-uniformity as low as 0.77% and a lag of 0.1%. The characterization showed that neither the dynamic range, the imager lag nor the PRNU are compromised with the proposed noise reduction technique. The characterization also shows that the QE of the PPD is barely impacted by the neighboring PMOS n-wells.
  Compared to recently reported CIS based on PPDs and featuring deep subelectron noise performance [1, 2], this readout chain presented in this work is faster, features a higher dynamic range and requires no process refinements and no input signal with a voltage higher than 3.3 V.

11.1.4 A New CMS Circuit Design for Further Noise Reduction

A new implementation of CMS based on a passive switch capacitor network has been presented. The proposed circuit topology allows performing a CMS of order 2^n using n additional capacitors and offers the advantages of being faster than classical digital CMS implementations since it requires a single analog-to-digital conversion, requiring no additional active circuitry and having no impact on the dynamic range of the readout chain. The newly proposed CMS circuit has been validated by transient noise simulation results confirming that the thermal noise reduction follows a $1/\sqrt{M}$ law and that the $1/f$ noise reduction reaches a plateau for CMS orders hier than 4.

11.1.5 *Technology Downscale: Lower 1/f and Thermal Noise but Higher Leakage Shot Noise*

The impact of the technology downscale on the read noise of CIS has been discussed based on the calculated input referred noise formulas and the evolution of the transistors process and geometrical parameters with the technology downscale. The analysis shows that the $1/f$ and thermal noise could be reduced dramatically to sub-$0.3\,e^-_{rms}$ level with technology nodes under 90 nm. But for such technology nodes, the gate leakage current for thin oxide transistors increases by several orders of magnitude compared to 180 or 130 nm technologies which might increase significantly the leakage current shot noise.

11.2 Read Noise Reduction in Terahertz CMOS Image Sensors

Another example of the readout noise reduction in image sensors circuit has been given in the field of CMOS THz imaging. CMOS THz detectors are based on metal antennas coupled with MOS transistor rectifiers. This front end transistor is the major noise contributor. In this case of active imaging, the THz source lightning the object can be controlled and offers an additional degree of freedom for noise reduction. Indeed, the noise reduction mechanism proposed in this work consists in modulating the THz source and performing in-pixel high selective filtering. This mechanism acts as chopper stabilized technique. The $1/f$ noise is reduced by shifting the signal above the corner frequency and thermal noise can be reduced by applying narrow band filtering. The filtering has been implemented at the pixel level. In this way, the integration time of several ms has been exploited by allowing high rise time of the order of several ms compatible with very high-Q filtering of 100. A 31 $\times$ 31 pixel focal plane array, for THz imaging, has been implemented in a 130 nm process and tested successfully. Each of the imager pixels comprise a bow tie antenna, a variable gain low noise amplification chain and a high-Q filter designed to be centered at modulation frequencies of the order of a few hundreds of kHz. The high-selective filter is based on a passive switched capacitor 16 paths filter optimized for low noise by a G_m-C filter. The N-path filter cycling frequency is generated on-chip with an external clock and locked to the modulation frequency of the THz source. In this way the center frequency of the narrow band filtering matches the modulation frequency. A simplified analytical model of the passive SC N-path filter has been proposed and good matching between the simulation, calculation and measurements has been shown.

The test and characterization of the THz imager showed that a Q of 100 have been reached with an input-referred noise of 0.2 μV. A drastic noise reduction of the noise PSD by 20 dB when applying the high-selective filtering has been demonstrated as expected theoretically. Thanks to its ultra low noise performance, this CMOS THz imager features a total NEP as low as 0.6 nW at 270 GHz. The imager has also been tested with different THz sources in the range of 200 to 600 GHz.

11.3 Future Work

11.3.1 CMOS Image Sensors

This works opens the way for the following future works:

- The experimental results demonstrated the validity of the noise analysis presented in this thesis. The sub-0.5 e^-_{rms} noise performance has been achieved using pixel-level optimization and column-level amplification [3, 4]. But the analytical and simulation results show that further noise reduction can be achieved by integrating the CMS circuit presented in Chap. 8. The integration of CMS would also allow reaching lower noise levels with lower column-level gain increasing the dynamic range of the readout chain in the high-gain mode.
- This thesis shows that the noise reduction at the pixel-level includes the source follower optimization and the sense node capacitance reduction. The image sensors presented in this thesis were designed using standard processes. Hence only the source follower optimization has been covered and the reduction of the sense node capacitance consisted in a careful layout. Different process optimizations reducing the sense node capacitance have been reported in recent works [1, 2, 5]. A combination between these process refinements and the source follower optimization is expected to further decrease the CIS read noise.
- The input-referred noise formulas calculated and validated in the frame of this thesis together with noise parameters data obtained from different design kits show that the read noise can be further reduced to levels below 0.3 e^-_{rms} by using technology nodes smaller than 180 nm. This has been confirmed with transient noise simulation results performed with design kits of advanced processes [6, 7].

11.3.2 Terahertz CMOS Image Sensors

The performance of the CMOS THz imager presented in this thesis can be further improved through a better antenna design with a higher responsivity and also by taking advantage of the technology downscale. Indeed, the noise reduction of the readout chain relies on the very high Q factor of the filter which is directly related to the area of the MOS capacitors. Thus, the design can also be improved by using a more advanced CMOS process with a higher oxide capacitance density that would allow the same selectivity for a smaller area in order to increase the fill factor. The fill factor of the presented pixel can also be improved by omitting the additional gain stages as well as the digital blocks included in the pixel.

References

1. J. Ma, D. Starkey, A. Rao, K. Odame, E. Fossum, Characterization of quanta image sensor pump-gate jots with deep sub-electron read noise. IEEE J. Electron Devices Soc. **3**(6), 472–480 (2015)
2. M. Seo, S. Kawahito, K. Kagawa, K. Yasutomi, A $0.27e^{-}_{rms}$ read noise 220 $\mu v/e^{-}$ conversion gain reset-gate-less cmos image sensor with $0.11 - \mu m$ cis process. IEEE Electron Device Lett. **36**(12), 1344–1347 (2015)
3. A. Boukhayma, A. Peizerat, C. Enz, Temporal readout noise analysis and reduction techniques for low-light cmos image sensors. IEEE Trans. Electron Devices **63**(1), 72–78 (2016)
4. A. Boukhayma, A. Peizerat, C. Enz, A sub-0.5 electron read noise vga image sensor in a standard cmos process. IEEE J. Solid-State Circuits **51**(9), 2180–2191 (2016)
5. K. Fumiaki, W. Shunichi, N. Satoshi, K. Rihito, S. Shigetoshi, Analysis and reduction of floating diffusion capacitance components of cmos image sensor for photon-countable sensitivity, in *International Image Sensors Workshop* (2015)
6. R. Capoccia, A. Boukhayma, C. Enz, Sub-electron cis noise analysis in 65 nm process, in *2016 IEEE International Conference on Electronics, Circuits and Systems (ICECS)* (2016), pp. 560–563
7. A. Boukhayma, A. Peizerat, C. Enz, Noise reduction techniques and scaling effects towards photon counting cmos image sensors. Sensors **16**(4) (2016), http://www.mdpi.com/1424-8220/16/4/514

allow the same sensitivity for a smaller area in order to increase the fill factor. Thus, fill factor of the presented pixel can also be improved by omitting [illegible] stages as well as [illegible] blocks included in the pixel).

References

1. [illegible]

[illegible]

Appendix A
Signal Variance Conservation with Sampling

The purpose of this appendix is to prove mathematically that the variance of a signal is conserved when the latter is sampled and held. This is to prove in particular that the sample and hold process performed during the CDS or CMS has no impact on the noise variance.

The signal or noise PSDs before and after the sample and hold process are denoted $S'(f)$ and $S(f)$, respectively. They are related by the following equation

$$\int_{-\infty}^{+\infty} S'(f) = \sum_{n=-\infty}^{+\infty} \int_{-\infty}^{+\infty} sinc^2(\pi f T_S) S(f - \frac{n}{T_S}). \tag{A.1}$$

The dirac trail represents the Fourier transform of the sampling process and the $sinc$ term represents the impact of the holding process. It is the Fourier transform of the the rectangular function taking 1 between 0 and T_s and 0 elsewhere. T_s is the sampling period.

The right term of (A.1) can be developed as

$$\sum_{n=-\infty}^{+\infty} \sum_{m=-\infty}^{+\infty} \int_{\frac{m}{T_S}}^{\frac{m+1}{T_S}} sinc^2(\pi f T_S) S(f - \frac{n}{T_S}). \tag{A.2}$$

By changing the variable of integration from f to $f + \frac{m}{T_S}$, the latter expression can be written as

$$\sum_{n=-\infty}^{+\infty} \sum_{m=-\infty}^{+\infty} \int_{0}^{\frac{1}{T_S}} sinc^2(\pi (f T_S + m)) S(f - \frac{n-m}{T_S}). \tag{A.3}$$

This expression can be rearranged as

A. Boukhayma, *Ultra Low Noise CMOS Image Sensors*,
Springer Theses, https://doi.org/10.1007/978-3-319-68774-2

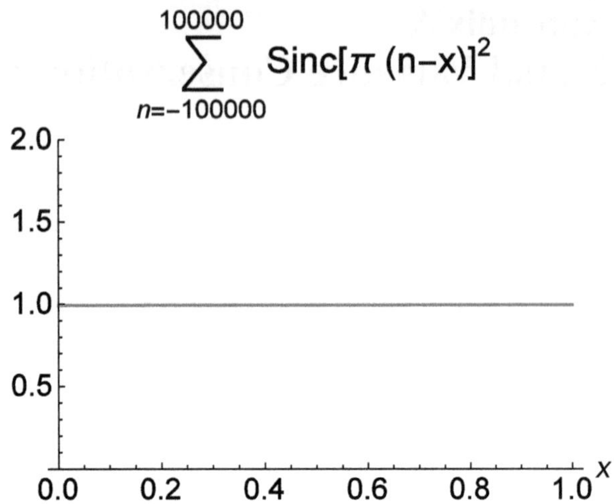

Fig. A.1 Numerical plot of $\sum_{m=-\infty}^{+\infty} sinc^2(\pi(fT_S+m))$

$$\int_0^{\frac{1}{T_S}} \sum_{m=-\infty}^{+\infty} \sum_{n=-\infty}^{+\infty} sinc^2(\pi(fT_S+m))S(f-\frac{n-m}{T_S})df. \tag{A.4}$$

By extracting the constant term in the second summation, the expression simplifies to

$$\int_0^{\frac{1}{T_S}} \sum_{m=-\infty}^{+\infty} sinc^2(\pi(fT_S+m)) \cdot \sum_{n=-\infty}^{+\infty} S(f-\frac{n}{T_S})df. \tag{A.5}$$

Now, we prove that the term $\sum_{m=-\infty}^{+\infty} sinc^2(\pi(fT_S+m))$ is simply equal to 1. The numerical evaluation of this sum, shown in Fig. A.1, confirms this hypothesis. Analytically, this sum can be expressed as

$$\sum_{m=-\infty}^{+\infty} sinc^2(\pi(fT_S+m)) = \frac{sin(\pi x)^2}{\pi^2} \sum_{m=-\infty}^{+\infty} \frac{1}{(x+n)^2}, \tag{A.6}$$

We use the summation (952) in [1]:

$$\sum_{m=-\infty}^{+\infty} \frac{1}{(n+\theta)^2+y^2} = \frac{\pi}{y} \frac{sinh(2\pi y)}{cosh(2\pi y)-cos(2\pi\theta)}, \tag{A.7}$$

Hence

$$\begin{aligned}\sum_{m=-\infty}^{+\infty} sinc^2(\pi(fT_S+m)) &= \frac{sin(\pi x)^2}{\pi^2}\sum_{m=-\infty}^{+\infty}\frac{1}{(x+n)^2}\\ &= \frac{sin(\pi x)^2}{\pi^2}\cdot\lim_{y\to\infty}2\pi^2\frac{sinh(2\pi y)}{2\pi y}\frac{1}{1-cos2\pi x}\\ &= \frac{sin(\pi x)^2}{\pi^2}\cdot\frac{2\pi^2}{1-cos2\pi x}\\ &= 1\end{aligned} \tag{A.8}$$

Consequently

$$\begin{aligned}\int_{-\infty}^{+\infty} S'(f) &= \int_0^{\frac{1}{T_S}}\sum_{n=-\infty}^{+\infty} S(f-\frac{n}{T_S})df\\ &= \sum_{n=-\infty}^{+\infty}\int_0^{\frac{1}{T_S}} S(f-\frac{n}{T_S})df\\ &= \sum_{n=-\infty}^{+\infty}\int_{\frac{n}{T_S}}^{\frac{n+1}{T_S}} S(f)df\\ &= \int_{-\infty}^{+\infty} S(f)df\end{aligned} \tag{A.9}$$

Reference

1. L. Jolley, *Summation of Series* (Dover Publications, New York, 1961), p. 176

Author Biography

Dr. Assim Boukhayma was born in Rabat, Morocco. He did the preparatory classes for graduate engineering school at Lycée Moulay Youcef, Rabat, Morocco, and was awarded with the excellence scholarship to pursue engineering studies in France. He received the graduate engineering degree (Diplôme d'Ingénieur) in telecommunications and the M.Sc. in microelectronics and embedded systems architecture from Institut Mines Telecom, Brest, France, in 2012. He was awarded with the graduate research fellowship for doctoral studies from the French atomic energy commission (CEA) and the French ministry of defense (DGA). He received the Ph.D. at the Swiss Federal Institute of Technology (EPFL), Lausanne, Switzerland, in the subject of Ultra Low Noise CMOS Image Sensors.

Assim is currently a research team leader in the Integrated Circuits Lab (ICLAB) headed by Professor Christian Enz at EPFL. His research focuses on image sensing, light sensors and integrated circuits. He worked between 2012 and 2015 as a doctoral researcher at CEA and DGA. Earlier in 2012 he worked for his M.Sc. thesis at CEA on the subject of THz CMOS Imaging. From 2010 to 2011, he worked with Bouygues-Telecom as a Telecommunication Radio Junior Engineer.

This book is the essence of Assim's PhD work between 2012 and 2016 that took place between Grenoble, France, and EPFL, Switzerland, with support from DGA.

Assim authored several patents and papers in scientific journal and conference proceedings. He gave several seminars in specialized workshops and at French

A. Boukhayma, *Ultra Low Noise CMOS Image Sensors*,
Springer Theses, https://doi.org/10.1007/978-3-319-68774-2

ministry of defense events. He has been invited to present this work in the "Forum de l'Innovation" at École Polytechnique, Paris. With Professor Christian Enz, they have been invited to give seminars in the Swiss IEEE chapter in ETH, Zurich, and EPFL.

MIX
Paper from responsible sources
FSC C105338

If you have any concerns about our products,
you can contact us on
ProductSafety@springernature.com

In case Publisher is established outside the EU,
the EU authorized representative is:
Springer Nature Customer Service Center GmbH
Europaplatz 3, 69115 Heidelberg, Germany

Printed by Libri Plureos GmbH
in Hamburg, Germany